Robson Alexsandro de Sousa
Virginia C. Lima Menezes
Claudivan F. de Lacerda

Salinity in string bean plants cv. Pitiúba

Robson Alexsandro de Sousa
Virginia C. Lima Menezes
Claudivan F. de Lacerda

Salinity in string bean plants cv. Pitiúba

Origin and consequences

ScienciaScripts

Imprint

Cover image: www.ingimage.com

This book is a translation from the original published under ISBN 978-3-330-76606-8.

Publisher:
Sciencia Scripts
is a trademark of
Dodo Books Indian Ocean Ltd. and OmniScriptum S.R.L publishing group

120 High Road, East Finchley, London, N2 9ED, United Kingdom
Str. Armeneasca 28/1, office 1, Chisinau MD-2012, Republic of Moldova, Europe
Managing Directors: Ieva Konstantinova, Victoria Ursu
info@omniscriptum.com

Printed at: see last page
ISBN: 978-620-8-53632-9

DEDICATE

To my parents, Romão Augusto de Sousa and Maria Ana de Oliveira Sousa, and to my aunt, Francisca Ana de Maria, for their love and understanding, which helped me to complete another stage in my academic life.

OFFER

To my children Romão Neto and Isabel and my beloved wife Virgínia.

CONTENTS

CHAPTER 1

INTRODUCTION

Considered a legume with a wide worldwide distribution, the cowpea *(Vigna unguiculata* (L.) Walp.), also known as mung bean, mung bean, cowpea, pinto bean, kidney bean and beach bean, has been widely used in human food since ancient times. Its presence is more concentrated mainly in the tropical regions of the globe, as these have similar soil and climate characteristics to its probable cradle of origin, Africa (MOUSINHO, 2005). Estimates from the FAO (2004) report that in 2003, around 9.82 million hectares of cowpeas were cultivated in the world, of which approximately 9.44 million hectares were cultivated on the African continent. World production in the same year was 3.72 million tonnes of dry beans, with an average yield of 378 kg $ha^{(-1)}$, with Nigeria standing out as the world's largest producer, with 2.1 million tonnes produced.

Due to the conditions of adaptability and the population's eating habits, the North and Northeast regions account for 95% to 100% of the total areas planted with string beans, in the states of Maranhão, Piauí, Ceará and Rio Grande do Norte (ARAÚJO and WATT, 1988; SENA and BISERRA, 1991). The state of Ceará is the country's largest producer of string beans, followed by Piauí. In the 2002/2003 harvest, the area planted in Ceará was 618,600 hectares, resulting in a production of 211,800 tonnes (FNP, 2004).

The crop adapts well to Brazilian conditions, especially those in the Northeast. The most suitable temperature for the development of string beans is between 20°C and 30°C. High temperatures during flowering can be harmful to the crop and reduce nodulation in the roots. Temperatures below 20°C can paralyse plant development (OLIVEIRA and CARVALHO, 1988). With regard to the physical and chemical characteristics of soils, it grows best in those rich in organic matter, with a pH close to 6.0, medium texture, deep, well-drained and of medium to high fertility. The water requirements of the crop, in order to obtain maximum yields, vary from 300 mm to 400 mm during its cycle, depending on the local soil and climate conditions (EMBRAPA, 2003).

According to Ayers and Westcot (1999) worldwide, the intensive use of good quality water has increasingly led to a decrease in its availability for new and old irrigation projects. In addition, the growing need to expand agricultural areas has led to the need to use water considered to be of inferior quality. As well as the re-utilisation of drainage water with high salt contents and the use of species capable of high profitability when irrigated with these types of water (RHOADES et aL, 2000).

In the north-east of Brazil, most of the water used for irrigation has a salt concentration in the range of 1 to 30 mmolc $L^{'1}$, corresponding to the electrical conductivity range of 0.1 to 3.0 dS nr^{1}, and its use in agriculture, combined with intense evaporation and the lack of adequate soil drainage, can cause salinity problems (HOLANDA and AMORIM, 1997). As a result, checking the suitability of water for irrigation purposes should mainly take into account factors related to the content and type of salts, soil characteristics, crop tolerance, climatic conditions, irrigation management and drainage conditions (RHOADES et aL, 2000).

The high concentration of salts is a stress factor for plants, as it has an osmotic activity that retains water, in addition to the action of ions on the protoplasm. Ayers and Westcot (1999) state that the ions that most often cause toxicity are chloride, sodium and boron. An excess of Na^+ and, above all, an excess of Cl^- in the protoplasm cause disturbances in the ionic balance (K^+ and Ca^{2+} in relation to Na^+), as well as the specific effect of ions on enzymes and membranes (LARCHER, 2000).

Therefore, the aim of this work was to assess the development of string beans when irrigated with water of different salinity levels and different ionic compositions, analysing the accumulation of salts in the plant; vegetative growth; biomass production; photosynthetic processes; proline, carbohydrate, N-aminosoluble and protein contents.

CHAPTER 2

LITERATURE REVIEW

2.1 Salinity problems in agriculture

2.1.1 Soil problems

The process of soil salinisation originates in its own formation, as it comes from the weathering of the parent rock, involving physical, chemical and biological processes through the action of climate, relief, living organisms and time. During weathering, the various constituents of rocks are released in the form of simple compounds. The sources of water-soluble salts are the primary minerals found in the soil and rocks of the earth's crust (RICHARDS, 1954; SANTOS, 2000).

The three main natural sources of salts in the soil are mineral weathering, atmospheric precipitation and fossil salts (those remaining from marine and lake environments). Salts can also be added to soils through human activities, including the use of irrigation water, highly saline brines or industrial waste (BOHN et al, 1985; QUEIROZ et. al., 1997). Although the main source of all salts found in the soil is the weathering of rocks, there are few examples where this source has directly caused soil salinity problems (RICHARDS, 1954).

Saline habitats are defined by the abnormally high content of readily soluble salts. Oceans, saline lakes and small bodies of water (ponds and pools) are aquatic saline habitats. Saline soils can be found on the continents in both arid and dry climates. In humid regions it is possible for soil to become saline due to the influence of mist or spray from the intertidal zone, as well as on dunes, lake shores and river floodplains (marshes). Salinised environments can also be found in the vicinity of saline deposits. In the form of aerosols, some ocean salts can be carried inland to a distance of 100 kilometres or more by winds and clouds (LARCHER, 2000).

The occurrence of saline soils is associated with natural causes or human activity. In the latter case, the use of irrigation that is incompatible with the physical, chemical and mineralogical characteristics of the soil profile has been identified as one of the main causes of soil salinisation. Furthermore, in arid and semi-arid regions, low rainfall and high evaporation rates make it difficult to leach salts, which accumulate in quantities that are harmful to plant growth, greatly aggravating the problem (FAGERIA and GHEYI, 1997; OLIVEIRA, 1997).

Salinity problems have also been associated with the water used for irrigation, poor drainage and the presence of sub-surface water rich in soluble salts at shallow depths. In these cases, i.e. when salinisation is due to anthropogenic action, it is known as secondary salinisation. In addition, salinisation can be caused by the application of fertilisers, excessively and piecemeal throughout the natural cycle, inducing the root system to osmotic stress (DIAS, 2004). Oliveira (1997) mentions that secondary salinisation is, in some cases, responsible for irreparable losses in the soil's productive capacity, rendering large tracts of cultivated land barren.

Human-induced salinity is more noticeable in environments with a high rate of potential evapotranspiration and low rainfall throughout the year. Induced salinity in these areas manifests itself as a result of the irrigation practised, and where control through drainage is not carried out or is carried out inadequately (OLIVEIRA, 1997).

In the north-east of Brazil, which is characterised by high evapotranspiration and low rainfall, the total area occupied by genetically saline soils (planosols, solonetz, solonchak and others) is estimated at over 9 million hectares (PEREIRA et al, 1985). In the semi-arid Northeast, the highest incidences of areas with secondary salinisation are concentrated in the lands most intensively cultivated with the use of irrigation in the so-called Irrigated Perimeters. Bernardo (1997) states that around 30% of the irrigated areas of public projects in north-eastern Brazil are suffering from salinisation problems.

There are basically two techniques for recovering soils affected by salts: washing off the salts and applying chemical improvements. Saline and/or sodic soils can rarely be reclaimed using just one technique. To achieve a successful reclamation process, two or more techniques are used. Washing is currently the most practical technique for eliminating salts from the soil, which consists of passing a certain amount of water through the soil profile capable of carrying away the salts present, which must be eliminated either superficially or underground through good drainage (SANTOS and HERNANDEZ, 1997).

2.1.2 Irrigation water quality

The quality of irrigation water can be considered an important factor, but it should never be forgotten that it is only one of the factors and that it is not possible to develop a universal classification system that can be used under all circumstances. The suitability of water for irrigation depends both on its own quality and on factors related to the conditions of use (AYERS and WESTCOT, 1999).

In recent decades, various classifications of water quality for irrigation have been established which, in general, have been lost because they are either too generalised and therefore not precise enough, or too restrictive and therefore cannot be extrapolated to other situations. In any case, these classifications commonly take into account (1) the effect of salinity (expressed as electrical conductivity, EC) on crop yields, and (2) the combined effects of salinity (EC) and sodicity (expressed as Sodium Adsorption Ratio, SAR) on the structural stability of soils. The toxic effect of some microelements and nutritional effects should be considered in certain circumstances (FAO, 1985 cited by ARAGÚES, 1994).

Bernardo et al. (2005) lists six basic parameters for analysing water and determining its quality for irrigation: a) total concentration of soluble salts or salinity; b) relative proportion of sodium in relation to other cations or soil infiltration capacity; c) concentration of toxic elements; d) concentration of bicarbonates; e) sanitary aspect; and f) aspect of emitter clogging (localised irrigation).

Salinity in irrigated areas occurs as a result of the use of water of inadequate quality, associated with the management of the soil-water-plant system and, whatever its source, the water used for irrigation always contains salts, although the quantity and quality of the salts present vary greatly (MEDEIROS and GHEYI, 1994).

Surface and groundwater originate from rainfall, which is the natural source of irrigation for rainfed farms. The salinity of rainwater is low, with an EC of around

0.010 dS rrr^{1} (MOLLE and CADIER, 1992), which favours the dispersion of soil particles and causes infiltration problems (AYERS and WESTCOT, 1999). Strictly speaking, all natural waters, whether rainwater, surface water (rivers, lakes and reservoirs) or underground water (wells and ditches), contain dissolved salts in varying quantities (HOLANDA and AMORIM, 1997). The natural quality of water results from the consequence of its origin and the natural physical-chemical processes in its various courses (LOPEZ, 2000).

According to Yaron (1973) quoted by Blanco (1999), the salt content of surface water is a function of the predominant rocks in the springs, the climate zone, the nature of the soil in which the water flows and pollution due to human activity. In the case of groundwater, the salt content depends on the origin of the water and the course over which it flows, in accordance with the law of dissolution, based on contact between the water and the substrate that stores it. Changes in the salt content of groundwater in the recharge process are the result of reduction (generally of a biochemical nature), cation exchange, evapotranspiration and precipitation.

Generally, the waters of humid zones have a lower salt content than those of arid zones, underground waters are saltier than river waters and the latter, close to the mouth, contain more salts than those close to the source; the concentration of salts in river waters varies with the season, being higher in autumn than in spring. In a given region, the waters naturally contain less salts before irrigation projects are developed (HOLANDA and AMORIM, 1997).

According to Hoorn (1971) and Kovda et al. (1973), evapotranspiration and rainfall are the main climatic elements to consider when assessing irrigation water. The amount of water to be applied during an irrigation period depends on evapotranspiration, which affects the irrigation regime and, consequently, the seasonal movement of salts in the soil profile. Greater evapotranspiration requires a greater quantity of irrigation water, which in turn leads to a greater quantity of salts and generally to greater soil salinity, although some of this is leached out (HOORN, 1971).

The problem of salinity in irrigated areas is aggravated when the salt balance shows a greater input than output, indicating a greater increase in salts in the area under consideration (BIGGAR et al, 1984).

The quality of irrigation water is related to its harmful effects on the soil and the crop, often requiring special management techniques to compensate for any problems associated with its use. The suitability of water for irrigation depends both on the quality itself and on factors related to the conditions of use. The same quality of water can be considered suitable for a certain crop or soil, but unsuitable for others. It is therefore essential to assess water quality as a preventative measure against gradual salinisation processes due to the accumulation of salts from successive irrigations. In addition to the physical-chemical characteristics of the water, other factors must be considered and analysed together when evaluating it and recommending its use for irrigation (AYERS and WESTCOT, 1999).

The main salts found in both soil and water are sodium chloride (NaCl), magnesium sulphate (MgSO4), sodium sulphate (Na2SO4), magnesium chloride (MgCh) and sodium carbonate (Na2COs). These salts can have their solubility affected by certain factors, such as pH and temperature at the source, which is why these factors must be taken into account when studying water quality, especially when carbonates and/or bicarbonates are present (ULZURRUN, 2000).

When determining the feasibility of using water for irrigation, the chemical composition of the water, crop tolerance, soil, water and crop management practices, climatological conditions, irrigation method and drainage conditions must be taken into account (CONTRERAS and ELIZONDO, 1980).

Waters from sedimentary regions with low salinity are mainly bicarbonate, sulphate or mixed, while in the fissures of the Precambrian rocks of the crystalline basement, which tend to be more saline, there is a predominance of only sodium and magnesium chloride waters (LEPRUN, 1983).

Silva Júnior et al. (1999), analysing waters from various sources in the states of Paraíba, Ceará and Rio Grande do Norte, concluded that the crystalline waters of north-eastern Brazil have low levels of sulphates, carbonates and potassium, as well as finding a predominance of sodic chloride waters regardless of the level of salinity.

Oliveira and Maia (1998), analysing 600 water samples from various aquifers of sedimentary origin in Rio Grande do Norte, showed that the greatest incidence of salinity problems, expressed by their high electrical conductivity and sodicity, occurred in samples from wells that exploit the resources of the Jandaíra limestone or aquifers supplied directly by the water table. They also showed that in the sedimentary area of Rio Grande do Norte, saline waters with EC > 3 dS rrr^{1} come proportionally from: tube wells 27%; Amazon wells (cacimbas) 16% and free springs (dams, rivers and streams) 21%.

Cruz and Melo (1969) state that in the sedimentary areas of north-eastern Brazil, groundwater has a low dissolved solids content, with concentrations predominantly below 500 mg L$^{'1}$. They also show that the flow gradients that determine the speed of recharge are factors responsible for the quality of the water, since the greater degree of mobility reduces mineralisation and the effects of concentration by evaporation, especially in the areas with the lowest depths that are susceptible to such effects.

Cruz and Melo (1974) cited by Silva Júnior et al. (1999), showed some factors responsible for the salinisation of groundwater in the crystalline region of north-eastern Brazil, including the process of concentration by evaporation, observing a direct relationship between the aridity index and the degree of salinisation of the water; thus the total concentration of salts varies according to the arid zoning, observing that the increase in concentration is accompanied by an increase in Cl^-, Na^+ and Mg^{2+}. Another factor that has also been reported is the condition of water circulation, where shallower fractures (below 40 m) intercommunicate with the alluvial fans, favouring the renewal of the aquifer, while in the deeper fractures circulation is more restricted, with more stagnant conditions, with greater exchange with the rock and thus the lithological influence overrides the climatic one.

2.2 Salinity problems for plants

2.2.1 How salts affect plants

Soil salinisation, or the accumulation of soluble salts in the root zone to levels that negatively affect crop yields, can be considered a natural process in arid or semi-arid systems, because potential evapotranspiration exceeds rainfall and limits the leaching of salts accumulated in the soil. This accumulation of salts occurs because the water is evapotranspired in its pure state, with the result that the dissolved salts are concentrated in the water remaining in the soil (ARAGÚES, 1995).

The main problems caused by soil salinisation are a reduction in the osmotic potential of the soil solution, reducing its water availability, and the toxicity of certain ions to plants (RHOADES et al., 2000; MUNNS, 2002; BERNARDO et al., 2005). The effects of salts on plants can result in water absorption difficulties for the plant, similar to those caused by excessive fertilisation, and can cause interference in physiological processes or even toxicity. Certain ions, such as Na^+ and Cl^-, can: (I) become toxic when present in the plant cell in excess; (II) damage the cell membrane; and (III) promote metabolic disturbances, affecting photosynthesis, respiration, protein metabolism, enzymatic activities, toxin production, etc. (MUNNS, 2002). According to Rhoades and Loveday (1990) and Bernardo et al. (2005), the effects of salinity are manifested in reduced germination and growth rates, reduced yields and, in certain severe cases, can cause widespread plant death.

Plants are classified according to salinity into halophytes and glycophytes. The variation in behaviour is very wide: from the most sensitive glycophytes, which don't grow normally even at very low sodium chloride contents, to the most resistant halophytes, such as *Rhizophora mangle*, which survives at salinities close to seawater, *Suaeda marítima* and *Salicornia europea* (LEVITT, 1980) or the halophytes of the Chenopodiaceae family, which have no growth problems at salinity levels of up to 15,000 mg L^{-1} of sodium chloride, although higher concentrations affect them negatively (LARCHER, 2000).

The first effect of salinity on plants can be described as the osmotic effect. The process of water extraction from the soil by plants takes place when the forces for water extraction by the root tissues are greater than the water retention forces exerted by the soil. As the moisture content of the soil decreases, the remaining water retention forces become greater. When the water in the soil is retained with a force greater than the extraction forces, a water shortage begins in the plant. The presence of salts in the soil solution causes the retention forces to increase, reducing the osmotic potential of the medium and, consequently, the magnitude of the plant's water shortage problem (AYERS and WESTCOT, 1999).

According to Ayers and Westcot (1999), plant toxicity problems arise when certain ions, soil or water constituents are absorbed by plants and accumulate in their tissues in concentrations high enough to cause damage and reduce yields. Toxicity problems arise from the high concentration of a specific cation or anion, as well as from an unfavourable saline composition in the soil solution, resulting in excessive or unbalanced absorption by plants (SMEDEMA and RYCROFT, 1988). The ions that most often cause toxicity are chloride (necrosis on the leaves), sodium (burns or necrosis along the edges of the leaves) and boron (dry spots on the edges and apex of older leaves) (AYERS and WESTCOT, 1999). However, Cruciani (1997) states that the effects of specific ions do not necessarily involve the direct action of the salt or ion on the plant tissue. Some are felt through the imbalance they cause in the absorption of essential nutrients in the plant's normal metabolism. Medeiros and Gheyi (1997) state that in many cases the concentration of salts does not reach osmotic levels capable of damaging the absorption of water by plants. However, the concentration of various ions can cause indirect interference and be an obstacle to the proper absorption of essential elements and, consequently, to the development of a normal metabolic process.

2.2.2 Crop tolerance to salinity

Salinity tolerance is a term that is not precisely defined due to the different conceptual approaches that can be taken.

Thus, tolerance has been defined as the degree to which a plant adjusts its osmotic potential with minimal sacrifice of its growth or as the measure of a plant's ability to withstand the effects of a saline solution concentrated in the root zone. Levitt (1980) associates tolerance with the absence of negative effects on the growth of plants that accumulate salts in their tissues. Salt tolerance is a protoplasmic component of resistance to salt stress. This tolerance involves the degree to which the protoplasm (depending on the plant species, tissue and vigour) can tolerate a disturbance in the ionic balance associated with salt stress, as well as its ability to tolerate toxicity and the osmotic effects of increased ion concentration (LARCHER, 2000).

Maas (1986) considers that tolerance can be analysed from three different points of view: (1) by the ability to survive in saline soils; (2) by absolute growth or production in saline soils and (3) through growth or production in a saline soil relative to a non-saline soil. Survival may be relevant from an ecological or conservation point of view, but it is a criterion of limited agronomic value, since survival is always accompanied by reductions in production to commercially unacceptable limits. On the other hand, survival mechanisms can be very different to those of production in moderately saline conditions. Thus, many halophytes maintain themselves in highly saline soils using various strategies such as dormancy, increased succulence or a shortened vegetative cycle (LEVITT, 1980). These survival mechanisms have little to do with production, since dormancy is incompatible with high yields and increased succulence does not lead to greater dry weight (ROYO and ARAGÚES, 1989).

The criterion of absolute production allows economic estimates to be made under salinity conditions, but limits comparisons between crops, since productions are not expressed in comparable terms. Absolute production under salinity conditions can be more indicative of vigour than actual tolerance (MAAS and HOFFMAN, 1977; NIEMAN and SHANNON, 1977).

Finally, the third criterion, relative yield, makes it possible to compare different crops whose yields are expressed in different units, making it the most appropriate criterion for judging the relative levels of salinity that both crops and their cultivars can tolerate (NIEMAN and SHANNON, 1977).

Some crops, due to their better osmotic adaptation capacity, are able to produce satisfactorily even in saline environments. This adaptability is very useful and allows for the selection of the most tolerant crops, capable of producing economically acceptable yields, when soil salinity cannot be kept below the tolerance level of traditionally cultivated plants. The maximum level of average salinity in the root zone that plants can tolerate without any negative effect on their development and yield is known as the threshold salinity (SL), which is usually expressed in terms of ECes (GHEYI et al., 1991; AYERS and WESTCOT, 1999).

Mass and Hoffman (1977) obtained a linear relationship between salinity and crop yields, represented by the equation Y = 100 - b (CEes - SL), where: Y corresponds to the percentage of potential yield; SL is the crop's threshold salinity level, in dS m^1; EC_{es} is the electrical conductivity of the soil's saturation extract, in dS m^1, expressing the soil's salinity condition; and "b", the potential yield loss caused by a unit increase above the crop's tolerance threshold electrical conductivity.

The plant's response to salinity does not depend solely on the concentration of salts. Therefore, a rigorous study of the plant's tolerance to salts should take into account, in addition to the amount of total soluble salts, other factors that often interfere with production, such as the plant, soil and climate (FAGERIA, 1989).

According to Dias (2004), the behaviour of plants in relation to salinity can vary according to their stage of development, although it is not clear whether this is due to susceptibility to salinity at a particular stage of growth or over the period in which the plant is exposed to the saline substrate, or the interaction between these factors.

As for the soil factor, according to Fageria (1989) and Ayers and Westcot (1999), crop tolerance can increase, decrease or even not be affected, depending on the fertility-salinity interactions that can occur in the soil. Maas and Hoffman (1977) add that reductions in salinity tolerance can occur as a result of excessive fertilisation, however, if the application of fertiliser does not cause a nutritional imbalance, there will be no significant effect on reducing crop tolerance to salts.

Finally, climate can affect the plant's response to salinity. In general, tolerance to salts is reduced in hot, dry climates (LACERDA, 1995; YEO, 1999) and is more pronounced in sensitive crops, whereas crops grown in cold climates or during the winter are more tolerant to salinity (MAAS, 1986).

Cheeseman (1988) states that, in the strict sense, tolerance mechanisms cannot be known, because tolerance in itself is a qualitative descriptor that reflects correlations between size, mortality and external salinity. Furthermore, the term salt stress is not interpretable at the level of mechanism because it is based on manipulations of an external environmental state that is only indirectly linked to the readjustment of cellular metabolism.

The physiological mechanism translates as osmotic adjustment, which is a process that results in a net accumulation of solutes in the cell, causing a decrease in water potential without a decrease in turgidity (TAIZ and ZEIGER, 2002). According to the same authors, osmotic adjustment can occur in both roots and leaves, with varying degrees of intensity. Fageria (1989) mentions that osmotic adjustment involves the absorption and accumulation of inorganic ions, as well as the synthesis of organic solutes such as carbohydrates and organic acids.

Many halophytes, under saline conditions, accumulate large quantities of ions in the leaves, which, however, are kept compartmentalised in the vacuoles of the cells, where they contribute to the osmotic potential without affecting the enzyme systems of the cytoplasm (TAIZ and ZEIGER, 2002). It should be emphasised that they often also accumulate amino acids, especially proline, as well as galactosyl-glycerol, organic acids and glycinabetaine; these organic compounds are of fundamental importance for osmotic adjustment in the cytosol (BUCHANAN et al., 2000).

In halophytic monocots, especially grasses, which contain less salt than halophytic dicots, the accumulation of soluble carbohydrates plays an important role, which is partly responsible for maintaining the low osmotic potential in the protoplast (LARCHER, 2000).

In glycophytes, osmotic adjustment also depends on the accumulation of synthesised organic substances (GREENWAY and MUNNS, 1980), leading to greater energy expenditure (YEO, 1981). According to Greenway and Munns (1980), glycophytes generally retain a large amount of salts in their stems and roots, so osmotic adjustment in the leaves depends more on the accumulation of organic solutes. The ability to accumulate organic solutes such as proline, glycinabetaine and soluble sugars, among others, is an important physiological characteristic that can favour plant growth in saline environments (SERRANO and GAXIOLA, 1994).

Most of the work to elucidate toxicity has centred on the ion Na^+ . Leopold and Willing (1984) found that the Na ion$^+$ caused an alteration in membrane permeability in soya beans. These same authors found that monovalent cations were more toxic than bivalent cations. Similar results were found by Ponnamperuma (1984) and Muhammed et al. (1987) in rice. Studies carried out by Reid and Smith (2000) point out the importance of Ca^{2+} in saline conditions, as a minimum level of this ion is necessary to maintain the integrity of membrane structures and the correct functioning of selective ion transport. Positive effects of Ca^{2+} have been found in beans (CACHORRO et al., 1994), sorghum (COLMER et al., 1994), barley (CRAMER et al.,1991; HUANG and REDMANN, 1995), alfalfa (CHAUNDARY et al., 1994), rice (ZHONG AND LAÚCHLI, 1994; LIN AND KAO.1995), maize (ALBERICO and CRAMER, 1993), tomatoes (SATTI et al., 1994), and various grasses (GALVEZ et al., 1993).

Because soil salinity is always associated with high concentrations of NaCl, many authors have focussed on finding relationships between salinity tolerance and the concentration of Na^+ and Cl' in the different parts of the plant and at different times during its growth stages. Thus, a relationship was found between Na^+ exclusion and tolerance in sorghum (MAAS et al., 1986; YANG et al., 1990; LACERDA et al., 2003), wheat (SALAMA et al., 1994), and several other species, including cowpea (COSTA et al., 2003).

2.2.3 String bean responses to salinity

String beans are a legume with a high protein value that is gaining greater agroeconomic value every day thanks to its adaptability to adverse conditions (CARVALHO et al., 2000). In the north-east of Brazil, it is a widely cultivated species, representing one of the staple foods of the population in this region (ARAÚJO and WATT, 1988; SENA and BISERRA, 1991), as well as having adaptive characteristics that are intrinsic to the north-east region, such as withstanding high temperatures and locations with water limitations (CALEGARI, 1995).

This species is considered to be moderately tolerant to salinity, and according to data obtained by the FAO, string beans can tolerate irrigation with saline water with an electrical conductivity of up to 3.3 dS m^{-1} and a soil electrical conductivity of 4.9 dS m^{-1} (AYERS and WESTCOT, 1999), without reducing productivity. However, this threshold salinity value can be variable, both due to the existence of intraspecific genetic variability in relation to this characteristic (COSTA et al., 2003), as well as depending on the type of soil, the climatic conditions of the region and the irrigation method used.

The concentrations of salts that restrict bean growth vary widely between varieties and depend not only on the type of salt, but also on the exposure time and the stage of development (SHANNON, 1992). Salinity due to high levels of Na^+ or Cl^- can inhibit the absorption of K^+ and Ca^{2+} ions by the roots, causing deficiency symptoms such as reduced growth, also affecting the content of amines as well as proteins (CRAMER et al., 1991).

Evaluating the tolerance of 50 cowpea genotypes under greenhouse conditions when subjected to different levels of salinity, Dantas et al. (2002) concluded that when salinity increased, there was a reduction in dry matter production in all the genotypes studied. The same authors also mention that the protein content in caupi grains as a function of soil salinity is independent of the genotype's tolerance to salinity in the growth phase. Soil salinity causes a reduction in plant protein levels, and both structural proteins and enzymes are affected. This reduction can cause both a slowdown in synthesis and an acceleration in degradation (DANTAS et al., 2002). LEMOS (1991), studying the cultivars IPA 202 and CNX 153-3f, when subjected to increasing levels of salinity in the soil, found a great influence of salt stress on the protein content of the grains.

Vasconcelos et al. (1987), evaluating the effect of different types of saline soils on the nodulation and growth of Pitiúba string bean plants, found that salinity affected nodulation and MVA infection of the roots, as well as the weight of dry matter in the aerial part. These authors found a strong inhibitory action of NaCl on plant growth, as well as albinism in the plants in the treatment with the highest level of salinity.

According to Silva et al. (2003), working with string bean plants grown in a nutrient solution under greenhouse conditions, showed that salt stress strongly inhibits growth, causes disturbances in the water balance and increases the total content of inorganic solutes in the different parts of the plants, an increase in the content of soluble carbohydrates in the leaves and roots, a reduction in the content of amino-soluble nitrogen, both at the root tips and in the younger trifoliate leaves, and also a reduction in the content of proline at the root tips.

Aragão et al. (2005) evaluated two cultivars of string beans, Vita 3 and Vita 5, one tolerant and the other susceptible to salinity, respectively, when subjected to a saline treatment with 0.01 M NaCl, and found that salinity had a greater influence on the Vita 5 cultivar, with a reduction in root and aerial part growth and a reduction in leaf area. However, chlorophyll content and fluorescence parameters were not affected by salt stress, but ribulose-1,5-biphosphate carbosylase/oxygenase (Rubisco) activity decreased in Vita 3 and increased in Vita 5.

In an attempt to minimise the effects of salinity on plants, some authors (LAHAYE and EPSTEIN, 1971; GRIEVE and MAAS, 1988) began to use Ca^{2+} supplements in the root environment of plants grown under conditions of saline stress. Positive results have been found (CRAMER et al, 1990; ORTIZ et al, 1994; LIN and KAO, 1995; COLMER et al, 1996), as well as the opposite, i.e. minimising the effects of salinity by adding supplementary calcium to the salinised environment (LEIDI et al, 1991; SONG and FUJYIAMA, 1996b). According to Lacerda (1995), these discrepancies in results may be due to differences between the species or cultivars used or the environmental conditions in which they were grown. This author found that the addition of calcium to the root environment of stressed plants of *Vigna unguiculata* cultivar Pitiúba, grown during the wet season, caused increases in the levels of ash, Na^+ and Cl' in the leaf limb compared to the saline treatment and that this aggravated the effects of salts on the growth and production of the crop. However, during the dry season, the use of Ca^{2+} reduced the Na^+ content in the leaf limb and minimised the deleterious effects of salinity on fruit production. Silva (1998) adds that the use of calcium in the root environment of stressed plants as an agent to minimise the effects of salinity is complex, as it depends on the source of calcium used, the plant species and the environmental conditions in which they are grown.

Guimarães (2005) studied the effects of applying different doses of calcium, in the form of the salts CaCh and CaSO4, to string bean plants cv. Pitiúba, subjected to salt stress, observed that salinity induced changes in gas exchange and water relations, causing reductions in net photosynthesis, transpiration rate and water and osmotic potentials and increases in leaf succulence, specific leaf weight and relative chlorophyll content, with these responses being little influenced by supplementary calcium. He also found that plant growth in saline medium (NaCl) was affected by both the source and concentration of calcium added to the medium. However, it was not possible to identify the limit dose of calcium in relation to vegetative growth in saline medium, since CaCh had a negative effect at concentrations above 2.5 mM and $CaSO_4$ had a positive effect up to the highest concentration used.

CHAPTER 3

MATERIAL AND METHODS

3.1 Location of the experiment

The experiment was carried out in the Vegetation House of the Biochemistry and Molecular Biology Department of the Federal University of Ceará - CE, located at 03°44' S and 38°33' W at an altitude of 19.5 m, from 5th May to 23rd June 2005.

3.2 Substrate

The substrate used was washed river sand, whose physical and chemical characteristics are shown in TABLE 1. The analyses were carried out at the Soil/Water Laboratory of the Department of Soil Sciences, Centre for Agricultural Sciences, Federal University of Ceará.

TABLE 1 - Physico-chemical analyses of the soil used in the experiment.

Features	
Textural Classification	SAND
Overall density (g cm^{-3})	1,54
Useful water (g $100g^{-1}$)	0,5
pH in water	7,03
EC (dS m^{1})	0,18
Ca^{2+} (cmoh kg^{-1})	0,50
Mg^{2+} (emole kg^{-1})	0,30
At^{+} ($cmolckg^{-1}$)	0,15
K^{+} ($cmolckg^{-1}$)	0,06
H^{+} + Al^{3+} ($cmolckg^{-1}$)	0,33
Al^{3+} ($cmolckg^{-1}$)	0,00
PST (%)	11,00

Source: Soil and Water Laboratory - DCS/CCA/UFC

3.3 Treatments and Experimental Design

Three types of water (A) with the following ionic compositions were studied: water 1 (A1), composed of salts of NaCl, $CaCl_2.2H_2O$ and MgCl2.6H2O, in a ratio of 7:2:1; water 2 (A2), made up of salts of Na2SO4, CaCh, MgCh, Na2COs, NaHCOand NaCl, in a ratio of 0.5 : 5: 1.4 : 0.2 : 1.5 : 1.4; and water 3 (A3) made up of salts of NaCl and $CaCl_2$, in a ratio of 9.96 : 0.04. TABLE 2 shows the chemical composition of the water used for irrigation. The waters were prepared according to the relationship between the ECa and its concentration ($mmol_c\,L^{-1}$ = EC x 10) taken from Rhoades et al. (2000). These waters were combined with 4 levels of electrical conductivity (C), i.e. 0.5 dS m^{-1}; 1.5 dS m^{-1}; 3.0 dS m^{-1} and 4.5 dS $m^{(-1)}$. The experimental design used was entirely randomised with 5 replications in a 3x4 factorial scheme, totalling 12 treatments listed in TABLE 3.

TABLE 2 - Chemical characteristics of the water used in the treatments.

Water	Ca^{2+}	Mg^{2+}	In^{+}	Cl^{-}	CO_3-	HCO_3-	SO_4-	EC	PH	RAS
	(mmolc L^{-1})							(dS m^{-1})		
Water 1	1,0	0,5	3,5	6,5	-	-	-	0,5	7,0	4,04
	3,0	1,5	10,5	19,5	-	-	-	1,5	7,3	7,00
	6,0	3,0	21,0	39,0	-	-	-	3,0	7,4	9,90
	9,0	4,5	31,5	58,5	-	-	-	4,5	7,3	12,12
Water 2	2,5	0,7	2,15	7,1	0,1	0,75	0,25	0,5	7,9	1,70
	7,5	2,1	6,45	21,3	0,3	2,25	0,75	1,5	8,2	2,94
	15,0	4,2	12,9	42,6	0,6	4,5	1,5	3,0	8,3	4,16
	22,5	6,3	19,35	63,9	0,9	6,75	2,25	4,5	8,4	5,10
Water 3	0,02	-	4,98	5,02	-	-	-	0,5	7,1	49,8
	0,06	-	14,94	15,06	-	-	-	1,5	7,0	86,3
	0,12	-	29,88	30,12	-	-	-	3,0	7,3	122,0
	0,18	-	44,82	45,18	-	-	-	4,5	7,2	149,4

Source: Soil and Water Laboratory - DCS/CCA/UFC

Water 1 = composed of NaCl, CaCfe.ZFkO and MgCh.SFkO salts; Water 2 = composed of Na2SO4, CaCh, MgCfe, Na2CO3, NaHCOs and NaCl salts; Water 3 = composed of NaCl and CaCb salts; EC = electrical conductivity of the water; RAS = sodium adsorption ratio.

3.4 Plant material

String bean seeds were used *(Vigna unguiculata* (L.) Walp.) cultivar Pitiúba, obtained from the Plant Physiology Laboratory of the Department of Molecular Biology and Biochemistry - UFC.

3.5 Setting up the experiment

Approximately 22kg of soil was placed in plastic pots with a diameter of 26cm and a height of 28cm, perforated on the underside. Before this, however, a 2 cm layer of gravel was placed to facilitate drainage. The pots were then placed on the central bench of the greenhouse (FIGURE 1), where the soil was washed three times with 2 litres of distilled water in order to remove the existing clay. Four seeds were planted in each pot.

FIGURE 1 - Arrangement of the pots in the greenhouse.

3.6 Irrigation

The watering shift was daily. The amount of water applied was estimated with the aim of the soil reaching its field capacity and the water starting to percolate, adding approximately a 15% leaching fraction, according to Ayers and Westcot (1999). Until germination, distilled water was used for irrigation.

3.7 Conducting the experiment

Germination took place 3 days after planting. Thinning was carried out 4 days later, leaving two plants per pot (FIGURE 2). After thinning, the treatments were applied. Fertilisation consisted of applying 200 mL per pot of a nutrient solution based on Hoagland's nutrient solution (EPSTEIN, 1975), the constituents of which are listed in TABLE 3. Ten applications were made throughout the crop cycle.

FIGURE 2 - View of the pot after thinning.

TABLE 3 - Composition of the nutrient solution used in the experiment.

Constituent	Quantity (mL L^{-1})
KNO_3 1M	6,0
$Ca(NO_3)_2$ 1M	4,0
NH4H2PO4 1M	1,0
MgSO4 0.5M	2,0
$(NH_4)_2$SO4	0,5
MICRO	1,0
Fe-EDTA	0,5

Source: Soil and Water Laboratory - DCS/CCA/UFC

The trial was monitored daily to prevent pest attacks and disease infestations. Two applications of Kumulus S, at a dose of 3 g L^{-1}, were made to control spider mites, as well as one application of Orthene (Acephate), at a dose of 1.5 g L^{-1}, to control leafhoppers.

The net photosynthetic rate of mature leaves was measured at 20 days after planting and 43 days after planting using an IRGA infrared gas analyser (LCA-

2, ADC, Hoddesdon, UK) in an open system with an air flow of 300 mL min^{-1} , as well as measuring transpiration rates and stomatal conductance using a porometer (LI-1600 Steady State Porometer, Li-Cor., Inc., Lincoln, Nebraska, USA). Measurements were always taken between 10am and 12pm, in full light, using relative humidity, temperature and ambient radiation.

At 47 days after planting, the chlorophyll content was measured using a Minolta SPAD-502 chlorophyll meter.

At 43 days and 48 days after planting, the electrical conductivity of the saturation extract was measured using a portable conductivity meter (Wet Sensor type Wet 1 CE, Delta T Devices, England).

The experiment was carried out 48 days after planting, at which time the leaf area was measured using a leaf area meter (LI - 3100, Area Meter, Li-Cor., Inc., Lincoln, Nebraska, USA) and the fresh weight of the leaves and stems was determined. The leaves and stems were placed in paper bags and dried in a forced circulation oven at 60 °C for 7 days, when the samples reached a constant weight, in order to obtain the dry matter of the aerial part. The specific leaf mass (mg cm^{-2}) and leaf succulence (g H2O dm^{-2}) were calculated according to Mantovani (1999).

When the experiment was carried out, three leaves from each plant were weighed and wrapped in aluminium foil and then placed in the fridge to be freeze-dried. This material was used to determine organic solutes.

After collecting the plants, the water drained from each pot was collected in order to carry out chemical analyses of the percolated water. The water was stored in plastic containers and placed in a refrigerator until the analyses were carried out. The root system was then collected, the roots washed and placed in paper bags to be placed in a forced circulation oven at 60 °C until a constant weight was obtained, in order to obtain the dry matter.

3.8 Determination of inorganic solutes

From the dry matter of the leaves and stems, which were finely ground in a Willey-type mill, the extract was obtained for the determination of Na content$^+$, Ca^{2+}, K^+ , P, Fe, Mn, Cu and Zn. The extract was prepared using 200 mg of the ground powder diluted in 15 mL of 1 N hydrochloric acid, then shaken for 60 minutes and filtered using rapid filter paper (MIYAZAWA et al., 1984). The Na^+ , Ca^{2+} and K^+ contents were determined using flame photometry, while the P contents were determined using colorimetry and the Fe, Mn, Cu and Zn contents using atomic absorption spectrophotometry (MALAVOLTA et al., 1989).

The Cl' contents were determined in aqueous extracts made by diluting approximately 100 mg of the dried and ground stem and leaf material with 10 mL of deionised distilled water. The homogenate was kept stirring for approximately 60 minutes and then filtered through filter paper. For the determination of Cl^-, an aliquot of 0.5 mL of the extract was used, topped up to 3.0 mL with demineralised water. Then 0.5 mL of a mixture of Hg(SCN)213.2mM in methanol and Fe(NO3)3 20.2% (4 + 1) was added, and the absorbance was read at 460 nm after 15 minutes. The Cl' levels were determined using a calibration curve with NaCl as the standard, prepared under the same conditions as above (GAINES et al, 1984). Each repetition was dosed by a single sample, without a replicate.

3.9 Determination of organic solutes

The extract was prepared from 100 mg of freeze-dried leaf powder, diluted in 10 mL of 0.01 M Na-K-Phosphate buffer solution with 0.01 M NaCl, pH 7.6; it was macerated for 10 minutes in a mortar and then the contents were transferred to test tubes and placed in a shaker for 30 minutes. They were placed in a centrifuge for 10 minutes at a speed of 3000 x g. Finally, the supernatant was filtered and stored in a freezer until it was analysed.

Carbohydrates were determined using 100 piL aliquots of the extract diluted in 400 µL of demineralised water, placed in test tubes; 500 µL of 5% phenol and 2.5 mL of concentrated sulphuric acid were added, then shaken in a vortex. It was left to cool and read by colorimetry at 490 nm. The concentration of carbohydrates was determined using a calibration curve with glucose as the standard (DUBOIS et al., 1956).

The protein content was determined by placing aliquots of 20 μL of the extract diluted in 80 μL of demineralised water in test tubes; 1 mL of Comassie was added, shaken with a vortex and a 15-minute reading was taken at 595 nm. The protein concentration was determined using a calibration curve, with albumin used as the standard (BRADFORD, 1976).

The procedure for the determination of N-aminosolubles consisted of using aliquots of 50 μL of extract diluted in 450 μL of buffer solution, adding 250 μL of T. Citrate, 500 μL of KCN and 100 μL of Ninhydrin; the reaction was heated in a water bath at 90 °C for 20 minutes; the reaction was stopped with an ice bath, then 650 pL of 60% ethanol was added. The reaction was stopped in an ice bath, then 650 pL of 60% ethanol was added. The colorimeter was read at 570 nm to find the concentration of amino acids using glycine as a standard using a calibration curve (YEMM and COCKING, 1955).

In the case of proline, 1 mL aliquots of the extract were placed in screw-top tubes, to which 1 mL of acid ninhydrin and 1 mL of glacial acetic acid were added. The mixture was shaken and placed in a water bath at 90 °C for 60 minutes. After this time, the tubes were cooled in an ice bath. After cooling, 2 mL of toluene was added to each tube and vortexed. The organic fraction (top) was removed using a Pasteur pipette and read at
520 nm. The concentration of proline was determined using a calibration curve with pure proline as the standard (BATES et al., 1973).

3.10 Statistical analyses

The results were subjected to analysis of variance and means were compared using the Tukey test at 5% probability, using the SAEG/UFV programme. Data from treatments with quantitative variables were subjected to regression analysis.

CHAPTER 4

RESULTS AND DISCUSSION

4.1 Analysis of percolated water and soil

TABLE 1 shows that the electrical conductivity of the drained water (ECad) increased considerably as the salinity of the water applied increased, reaching values that indicate a high risk of salinisation (Ayers and Westcot, 1999).

The increase in ECd shows the high rate of percolation, mainly due to the texture of the substrate used in the experiment (TABLE 1). The leaching process, according to Gheyi et al. (1991), is an effective way of controlling the gradual increase in soluble salts in the root zone and the extra amount of water applied percolates below the root zone, removing at least part of the accumulated salts.

Also in TABLE 1, we see that the composition of the drainage water is greatly influenced by the composition of the irrigation water (TABLE 2), since in the treatments where irrigation water with 50 % CaCh was used (T5, T6, T7 and T8), the drainage water has higher levels of $Ca^{(2+)}$ when compared to the other treatments where the presence of calcium salts in the irrigation water is lower. It can also be seen that the sodium and chloride values in the drainage water are quite high in all the treatments, with no difference between the initial composition of the irrigation water.

These results show what Gheyi et al. (1991) say, that not all the salts incorporated by the water remain in the soil, even in arid and semi-arid regions. However, these same authors state that assessing the risk of salinisation in a given area through the composition of irrigation water cannot be done simply by looking at the quality of the water, as its effects depend on the soil, the climate and the user's skill and knowledge in managing the water-soil-plant system.

TABLE 1 - Chemical characteristics of the water drained from the treatments and collected at the end of the experiment.

Treatments		Ca^{2+}	Mg^{2*}	In^{+}	K^{+}	Cl-	HCO3 -		PH	
WATER	EC_a (dS m^{-1})	(mmolc L'1)								
Water 1	0,5	4,1	2,70	28,5	0,10	35,20	0,40	2,93	6,90	15,46
Water 1	1,0	8,0	6,20	61,7	0,10	75,60	0,50	5,66	6,60	23,15
Water 1	3,0	8,7	5,00	53,0	0,10	65,20	0,80	5,66	7,04	20,25
Water 1	4,5	9,8	6,20	61,7	0,10	76,80	0,70	6,29	7,07	21,81
Water 2	0,5	5,2	4,00	15,0	0,10	24,40	0,50	2,32	6,67	6,99
Water 2	1,0	18,4	8,10	38,6	0,10	64,80	0,90	5,39	6,84	10,60
Water 2	3,0	22,2	9,10	43,9	0,10	73,60	2,00	6,26	7,64	11,10
Water 2	4,5	21,5	14,00	35,3	0,10	69,60	0,80	6,11	7,82	8,38
Water 3	0,5	2,00	2,80	33,5	0,10	38,10	0,80	3,05	6,84	21,62
Water 3	1,0	1,40	2,40	47,7	0,10	50,20	0,90	3,96	7,16	34,60
Water 3	3,0	1,20	1,30	76,0	0,10	77,20	1,10	5,59	7,20	67,98
Water 3	4,5	1,60	1,20	59,9	0,10	61,60	1,30	5,84	7,90	50,62

Source: Soil and Water Laboratory - DCS/CCA/UFC
Water 1 = composed of salts of NaCl, CaCĺ2.2H_2O and $MgCl_{(2)}$.6H_2O; Water 2 = composed of salts of Na2SO4, CaCh, $MgCl_2$, Na2CO3, NaHCOs and NaCl; Water 3 = composed of salts of NaCl and $CaCl_2$ EC_a = electrical conductivity of irrigation water EC_ad = electrical conductivity of drained water; RAS = sodium adsorption ratio

Analysing the soil's electrical conductivity (ECes), FIGURE 3, we can see an increase towards the end of the experiment. Although the EC values were much higher than the initial values, the results found are within the parameters of the classification of normal soils proposed by Richards (1954), which is partly due to the sandy texture of the substrate and the application of the leaching slurry. However, the use of water with a low or high salt content for irrigation purposes should always be monitored, as the accumulation of salts in the soil, even if gradual, can cause serious problems for the soil and, consequently, for agricultural production. According to Dias (2004) and Oliveira (1997), the process of secondary soil salinisation is responsible for irreparable losses in soils, significantly affecting crops.

Increasing levels of salinity in the irrigation water strongly influenced all the parameters assessed, while the interaction between the ionic composition of water and salinity was only significant for the parameters leaf dry weight and total dry matter (TABLE 2).

TABLE 2 - Mean square values and statistical significance for leaf dry weight (LSW), stem dry weight (SCW), root dry weight (RDSW), aerial part dry weight (APW), total dry matter (TDM), root/ aerial part ratio (R/AP) and total leaf area (TLA) in cowpea plants *(Vigna unguiculata* (L.) Walp.) Walp.) cv. Pitiúba, irrigated with saline water with different ionic compositions.

Sources of Variation	Mean Square						
	PSF	PSC	PSRAIZ	PSPA	MSTOT	R/PA	AFT
Water	3.168^{ns}	9,870*	5,950*	3.706^{ns}	5.322^{ns}	0,014**	1645333*
Salt	79,932*	79,756*	31,838*	316,922*	542,288*	0,014**	10928200*
Water x Salt	4,297*	1.245^{ns}	1.230^{ns}	6.662^{ns}	11,167*	0.0013^{ns}	73005.29^{ns}
Waste	1,633	0,965	0,978	3,445	4,694	0,0023	86603,62
C.V. (%)	13,61	8,21	23,85	8,69	8,49	25,30	7,87

'Significant by F-test at 5%; ** Significant by F-test at 1%; ns = not significant

TABLE 3 shows that the highest average stem dry weight was obtained using Water 3 (rich in sodium chloride) and Water 1 (rich in calcium and magnesium salts), while Water 2 (rich in calcium and bicarbonates) had the lowest value for this variable. On the other hand, the dry weight of the root was higher in the treatments irrigated with Water 2, with higher values than those irrigated with Waters 1 and 3, which resulted in higher values for the root/aerial part ratio when water rich in bicarbonates (A2) was used. The highest leaf area values were also observed in the treatments that used water rich in calcium and bicarbonates (A2), showing a much higher average value than when Water 1 and Water 3 were used.

TABLE 3 - Average contents of stem dry weight (SCW), root dry weight (RRW), root/aerial part ratio (R/AP) and total leaf area (TLA) in cowpea plants *(Vigna unguiculata* (L.) Walp.) cv. Pitiúba, when irrigated with water of different ionic compositions. Averages followed by the same letter in the columns do not differ by the Tukey test at 5% probability.

Treatments	PSC (g)	PSRAIZ (g)	R/PA	AFT (cm^2)
Water 1	12,36 a	4.18 ab	0.1912 ab	3683,8 b
Water 2	11,16b	4,68 a	0,2187 a	4047,2 a
Water 3	12,39 a	3,59 b	0,1640 b	3481,1 b

Water 1 = composed of salts of NaCl, CaCl2.2H2O and MgCl2.6H2O; Water 2 = composed of salts of Na2SO4, CaCl2, MgCl2, Na2CO3, NaHCO3 and NaCl; Water 3 = composed of salts of NaCl and CaCl2

The dry weight of the leaves, stem, root and aerial part, as well as the total dry matter of the string bean plants showed a decreasing linear behaviour (FIGURES 4A, 4B, 4C, 4D and 4E) as the water salt levels increased. Similar results were found in barley by Cramer et al. (1991), and in string bean plants subjected to increasing levels of NaCl (LACERDA, 1995; SILVA, 1998; COSTA et al., 2003). Dantas et al. (2002) found that when salinity increased in string bean genotypes, there was a reduction in dry matter production. The reduction in the biomass of the cowpea is scientifically supported by various studies which show that the relative water content in the leaves of cowpea plants under salinity remains almost unchanged, indicating an efficient protective mechanism coupled with the closure of the stomata (DANTAS et al., 2002). Larcher (2000) states that biomass production is a good criterion for assessing salt stress in plants.

It can be seen in FIGURE 4F that in this experiment the root/shoot ratio decreased as the salinity of the irrigation water increased, as a result of the greater effects of salinity on the growth of the root system. There is a decreasing linear behaviour when using Waters 2 and 3, and a quadratic behaviour when using Water 1, showing a slight tendency towards a higher root/shoot ratio, but with increasing salinity, there is a reduction in this variable. There was a 20 per cent reduction in relation to the lowest level of irrigation water salinity (ECa = 0.5 dS $m^{-1)}$. Although root growth is sensitive to salts (CRAMER et al., 1989), most studies show that the reduction in growth is greater in the area than in the roots (GREENWAY and MUNNS, 1980). It is possible that the growing conditions influence this variable.

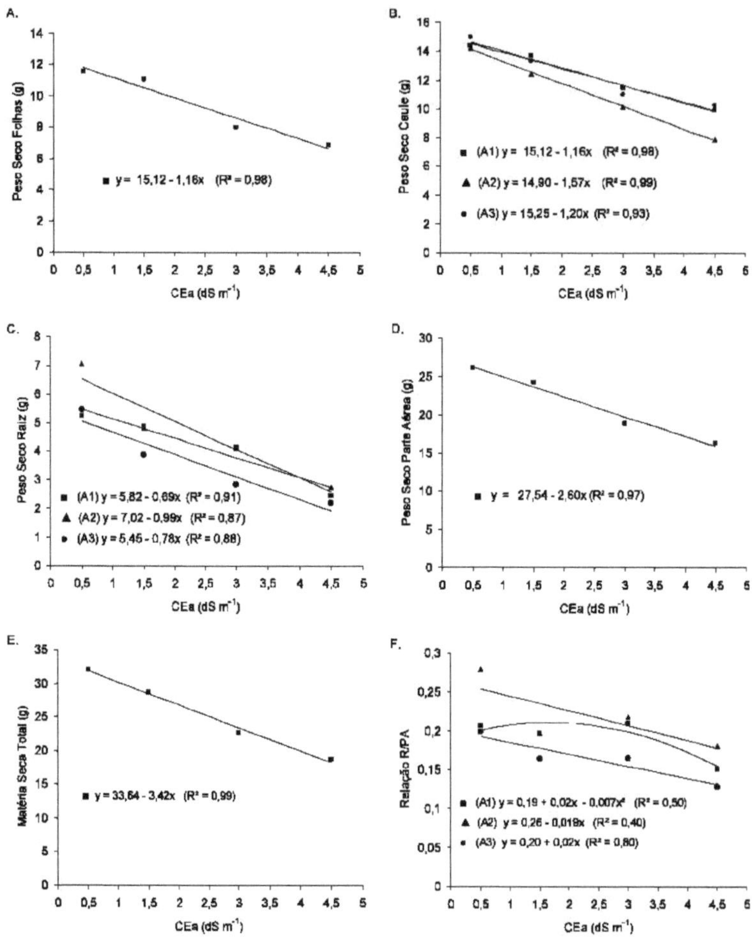

FIGURE 4 - Leaf dry weight (A), stem dry weight (B), root dry weight (C), aerial part dry weight (D), total dry matter (E) and root/ aerial part ratio (F) of cowpea plants *(Vigna unguiculata* (L.) Walp.) cv. Pitiúba as a function of irrigation water salinity.

An examination of the effect of salinity on leaf area is shown in FIGURE 5, which shows a linear decrease in leaf area as the salinity of the irrigation water increases. One of the first responses of plants subjected to saline stress is a reduction in the rate of leaf growth, reducing the leaf area available for photosynthesis (TERRY and WALDRON, 1984). This reduction in leaf area was also confirmed by O'Leary (1971) and Sousa (1995) in *Phaseolus vulgaris plants,* in cowpea plants by Silva (1998), Costa et al. (2003) and Aragão et al. (2005) and in banana plants by Carmo et al. (2003).

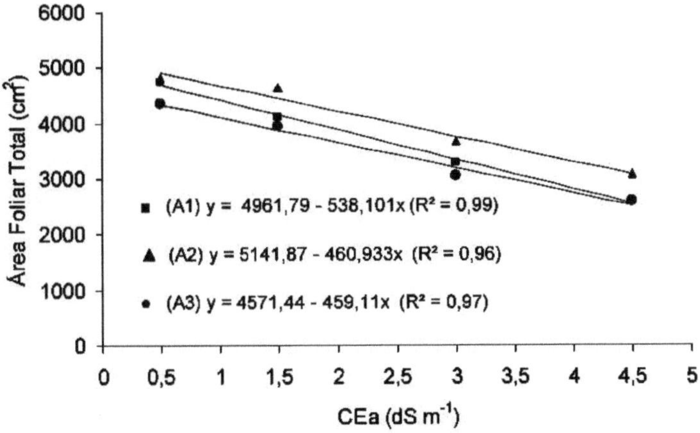

FIGURE 5 - Total leaf area of cowpea plants *(Vigna unguiculata* (L.) Walp.) cv. Pitiúba as a function of irrigation water salinity.

4.3 Gas exchange and morphophysiological characteristics

TABLE 4 shows that the transpiration rate and stomatal conductance were negatively influenced by the increasing effect of salinity in cowpea plants; however, for water composition and the interaction, both transpiration and stomatal conductance were not statistically significant. The net photosynthetic rate was not statistically significant for any of the sources of variation.

TABLE 4 - Mean square values and statistical significance of transpiration rate (E), stomatal conductance (GS), photosynthetic rate (FOT), leaf succulence (SUC), leaf specific mass (MEF) and chlorophyll (OLOR) in cowpea plants *(Vigna unguiculata* (L.) Walp.) cv. Pitiúba, irrigated with saline water of different ionic compositions.

Sources of Variation	Mean Square					
	E	GS	FOT	SUC	MEF	CLOR
Water	11.45^{ns}	0.012^{ns}	26.03^{ns}	0,1407*	0.145^{ns}	198,66**
Salt	17,05*	0,088*	2.34^{ns}	0,3281*	0.123^{ns}	24.00^{ns}

Water x Salt	3.61^{ns}	0.012^{ns}	6.66^{ns}	0,1814*	0.278^{ns}	14.46^{ns}
Waste	4,19	0,026	9,98	0,0200	0,153	10,34
C.V. (%)	13,22	15,28	16,02	8,47	15,57	6,66

'Significant by F-test at 5%; ** Significant by F-test at 1%; ns = not significant.

The effect of salinity on the variables transpiration rate and stomatal conductance can be seen in FIGURE 6. It can be seen that the effect of salinity on gas exchange was much less pronounced than on leaf growth, with the highest concentration of salts in the irrigation water showing an average reduction of 17% in stomatal conductance (FIGURE 6A) and 12% in transpiration rate (FIGURE 6B). Discrepancies between the effects of salinity on leaf growth and gas exchange have been observed by other authors (GUILHERME et al., 2003), indicating that the reduction in water consumption in plants grown in a saline environment may be more associated with a reduction in total leaf area than a reduction in the transpiration rate per unit leaf area.

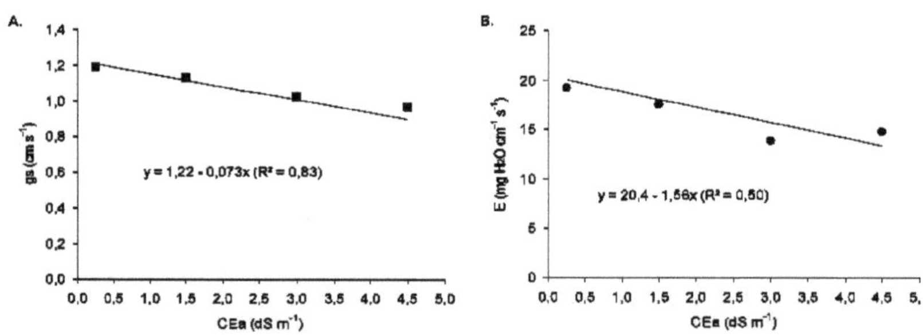

FIGURE 6 - Stomatal conductance (A) and transpiration rate (B) in leaves of string bean plants as a function of irrigation water salinity levels (ECa).

Reductions in transpiration rate due to salinity were also found by Silva et al. (2003), Azevedo Neto et al. (2004) and Guimarães (2005). Silva (1998) found that stomatal conductance and transpiration rate also decreased as salinity increased in cowpea plants. The reduction in stomatal conductance and transpiration may be associated with the partial closure of stomata due to the reduction in hydraulic conductivity of the root system caused by salinity, leading to an increase in suberisation and lignification of the vascular tissues of the roots (AZAIZEH et al., 1992; CACHORRO et al., 1993; SONG AND FUJIYAMA, 1996a, 1996b; PEYRANO et al, 1997; SILVA, 1998; GUIMARÃES, 2005), causing a reduction in the flow of water through the plants and, consequently, a reduction in the rates of stomatal conductance and transpiration (SILVA, 1998).According to Amorim and Valle (1993), the process of transpiration controls the absorption and translocation of water and mineral salts and is an important factor in the dissipation of energy as a form of cooling.When plants absorb CO2, they lose water through the leaves, mainly through the stomata, which have mechanisms to control their degree of opening; this mechanism is attributed to leaf stomatal conductance, a parameter used as an indicator of water deficiency (Mc DERMIT, 1990).

According to the values shown in TABLE 4, we can see that there was a significant difference for the leaf juiciness and chlorophyll variables when using irrigation water with different ionic compositions; however, the specific leaf mass was not statistically significant. Leaf juiciness showed a significant behaviour at the 5% probability level when salts were added to the irrigation water, as well as for the interaction between the water's ionic composition and salt levels, which was not the case for specific leaf mass and chlorophyll (TABLE 4). TABLE 5 shows the average values and the respective standard deviation for specific leaf mass and chlorophyll.

TABLE 5 - Means and standard deviation of chlorophyll (CLOR), in arbitrary units, and leaf specific mass (MEF), in mg cm^{-2}, in Pitiúba cowpea plants irrigated with water of different ionic composition at increasing salt levels.

Parameters	ECa (dS m^{-1})			
	0,5	1,5	3,0	4,5
CHLOR Water 1	46,46 ±4,17	48,06 ± 2,63	50,046 ± 2,56	51,74 ±2,07
CHLOR Water 2	46,16 ±4,71	44,04 ±3,12	44,44 ± 2,26	44,64 ± 3,52
CHLOR Water 3	49,84 ± 1,50	50,44 ± 2,68	49,54 ± 3,22	54,08 ± 4,45
MEF Water 1	2,36 ± 0,25	2,73 ±0,13	2,70 ± 0,38	2,50 ± 0,23
MEF Water 2	2,35 ± 0,24	2,58 ± 0,20	2,38 ± 0,24	2,36 ± 0,30
MEF Water 3	2,86 ± 0,49	2,57 ± 0,23	2,13 ± 1,01	2,66 ±0,17

Water 1 = composed of salts of NaCl, CaCl2.2H2O and MgCl2.6H2O; Water 2 = composed of salts of Na2SO4, CaCl2, MgCl2, Na2CO3, NaHCO3 and NaCl; Water 3 = composed of salts of NaCl and CaCl2

The leaf juiciness variable obtained the highest average values when Water 2 was used, whose ionic composition predominates calcium salts, carbonates and bicarbonates, followed by Water 3 and Water 1, the latter of which did not differ statistically (TABLE 6). The chlorophyll variable, on the other hand, behaved differently, as Waters 3 and 1 had the highest average values, surpassing Water 2 (TABLE 6).

TABLE 6 - Average contents of leaf succulence (SUC), in g H2O dm^{-2}, and chlorophyll (CLOR), in arbitrary units, in cowpea plants *(Vigna unguiculata* (L.) Walp.) cv. Pitiúba, when irrigated with water of different ionic compositions. Averages followed by the same letter in the columns do not differ by the Tukey test at 5% probability.

Treatments	SUC	CLOR
Water 1	1,60 b	49,08 a
Water 2	1,77 a	44,82 b
Water 3	1,64 b	50,98 a

Water 1 = composed of salts of NaCl, CaCl2.2H2O and MgCl2.6H2O; Water 2 = composed of salts of Na2SO4, CaCl2, MgCl2, Na2CO3, NaHCO3 and NaCl; Water 3 = composed of salts of NaCl and CaCl2

In FIGURE 7, we can see that, when subjected to saline stress, string bean plants showed an increase in leaf succulence in relation to the lower levels of salinity in the irrigation water, explained by quadratic and linear models, with the greatest increases being observed in the saline treatments with water 2, which contained the highest concentration of chloride. Similar results were also observed by Silva (1998), Costa et al. (2003) and Guimarães (2005) when they assessed salt-stressed Pitiúba cowpea plants. These results are also in line with those obtained by other authors who showed that succulence is a common response in the leaves of plants stressed with chloride salts, although it is more common in dicotyledons (STROGONOV, 1964; ZEKRI, 1991).

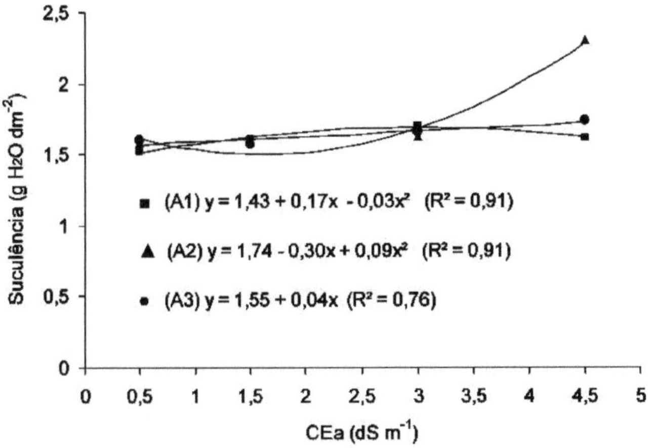

FIGURE 7 - Leaf juiciness of cowpea plants *(Vigna unguiculata* (L.) Walp.) cv. Pitiúba as a function of irrigation water salinity.

4.4 Mineral composition

4.4.1 $Ca^{(2+)}$> K^+ and P

The absorption and localisation of Na^+ , Cl^-, K^+ and Ca^{2+} in plants subjected to saline stress has received a lot of attention from researchers who want to study the mechanisms of tolerance to salinity (GREENWAY and MUNNS, 1980; LACERDA, 2005). Analysing TABLE 7, we can see that salinity had a significant effect on the calcium and phosphorus levels in the leaves and stems of Pitiúba string bean plants, at the 1% probability level, and only potassium was not statistically significant; while the ionic composition of the irrigation water influenced all the ions analysed except phosphorus, both in the leaves and stems (TABLE 7). As for the water x salt interaction, only the potassium ion variable in the stem showed statistical significance at the 5% probability level (TABLE 7), while in the leaves only the water x salt interaction for the calcium ion variable was statistically significant at the 1% probability level (TABLE 7).

TABLE 7 - Mean square values and statistical significance for the variables calcium (Ca^{2+}), potassium (K^+) and phosphorus (P) in the stem and leaves of cowpea plants *(Vigna unguiculata* (L.) Walp.) cv. Pitiúba, irrigated with saline water of different ionic compositions.

Sources of Variation	Mean Square					
	Stem			Leaves		
	Ca^{2+}	K^+	P	Ca^{2+}	K^+	P
Water	20,83**	31,560**	1.253^{ns}	22,754**	18,479*	1.358^{ns}
Salt	10,33**	10.577^{ns}	10,318**	2,813**	7.851^{ns}	8,917**
Water x Salt	0.549^{ns}	11,083*	0.593^{ns}	0,864**	9.244^{ns}	0.482^{ns}
Waste	0,569	3,835	0,687	0,131	4,295	0,682
C.V. (%)	8,10	8,85	18,35	15,98	9,63	18,80

'Significant by F-test at 5%; ** Significant by F-test at 1%; ns = not significant

The calcium ion responses in string bean plants when subjected to increasing levels of salts in the irrigation water are shown in FIGURE 8. There is an increase in calcium ion levels in the leaves, even when using different ionic compositions of water, while there is a decrease in the stem, results that are different to those observed by various authors (KENT and LÃUCHLI, 1985; GRIEVE and FUJIYAMA, 1987; GRIEVE and MAAS, 1988; CRAMER et al, 1989; NAKAMURA et al., 1990; SILVA, 1998). However, we can see that the calcium content in the leaves is higher than in the stem (FIGURE 9), a result also found by Lacerda (1995) and Silva (1998) in Pitiúba cowpea plants subjected to salt stress.

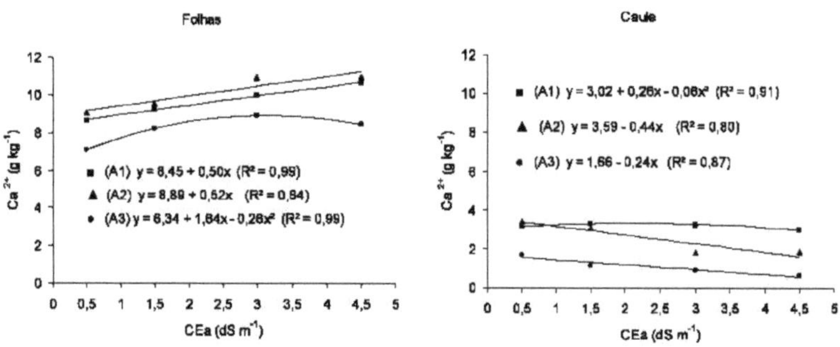

FIGURE 8 - Calcium content in the leaves and stem of string bean plants *(Vigna unguiculata* (L.) Walp.) cv. Pitiúba as a function of irrigation water salinity.

Calcium levels, according to the chemical composition of the irrigation water (TABLE 8), were highest in the leaves when irrigation was carried out with water 2 (predominance of CaCh) and water 1 (presence of CaCh) and lowest with water 3 (no calcium). The same result can be observed for the stems of the plants, since the accumulation of calcium ions was influenced, in decreasing order, by waters 1, 2 and 3. The increase in the calcium content of the cowpea leaves in this experiment can be explained by the plant's greater absorption of this ion, since the composition of the irrigation water had a predominance of calcium salts, except for water 3.

Several studies have demonstrated the effects of calcium as a reducer of sodium levels in maize (MAAS and GRIEVE, 1987; ALBERICO and CRAMER, 1993), beans (CACHORRO et aL, 1994), sorghum (COLMER et aL, 1994; LACERDA et al., 2003), barley (CRAMER et al., 1991; HUANG and REDMANN, 1995), alfalfa (CHAUNDARY et al., 1994), rice (ZHONG and LAUCHLI, 1994; LIN and KAO,1995), tomatoes (SATTI et al., 1994), and various grasses (GALVEZ et al., 1993). The beneficial effect of calcium in preventing the entry of sodium occurs primarily in the plasmalemma of the root cells involved in absorption. Studies carried out by Reid and Smith (2000) point out the importance of Ca^{2+} in saline conditions, since a minimum level of this ion is necessary to maintain the integrity of the membrane structure and the correct functioning of selective ion transport. On the other hand, the antagonistic effect of excess calcium on the absorption of other ions, such as potassium, is well known (MARSCHNER, 1995). Therefore, different plant responses to the salinity of the environment are possible when comparing different types of salts. These differences can occur in terms of growth, mineral nutrition or other physiological responses of the plant.

The accumulation of phosphorus in Pitiúba cv string bean plants was greatly influenced by salt stress, as can be seen in FIGURE 9. There was an increase in the content of this ion in both the leaves and the stem for the three types of water (A1, A2 and A3), a result similar to that found by several authors (LEVITT, 1980; MAAS and GRIEVE, 1987; ANDRADE, 1989; LACERDA, 1995; AZEVEDO NETO, 1997; SILVA, 1998). These results highlight the need for caution when carrying out phosphate fertilisation in saline environments, as the optimum level of this nutrient for plant production in the absence of salts can be toxic for some plants grown in saline environments (NIEMAM and CLARK, 1976; GRATTAN and MAAS, 1984). FIGURE 9 shows that the phosphorus ion is well distributed in the aerial part of the plant, even when there is an increase in the levels of salts in the irrigation water.

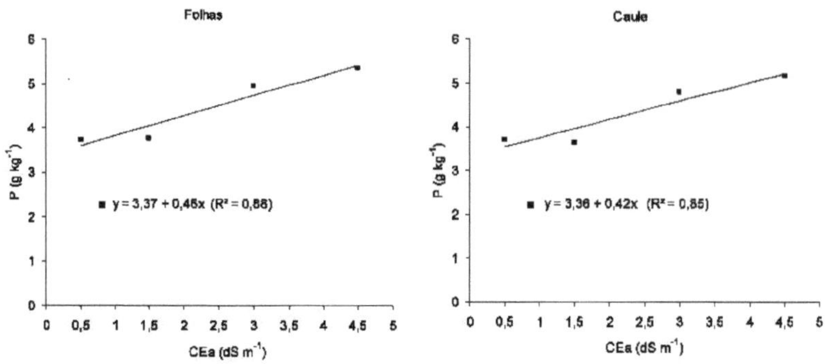

FIGURE 9 - Phosphorus content in the leaves and stem of string bean plants *(Vigna unguiculata* (L.) Walp.) cv. Pitiúba as a function of irrigation water salinity.

On the other hand, some authors have argued that the increase in P absorption and accumulation in stressed plants is more common when P concentrations are high in the growing solution (ROBERTS et al., 1984; NAVARRO et al., 2001), as seen in this study, and may not occur in plants grown in the field or in a medium poor in this nutrient. This observation was confirmed in a study carried out with two forage sorghum genotypes grown under greenhouse conditions in pots containing sandy soil and irrigated with saline water of up to 8 dS m-1 (AQUINO, 2005). This author found no trend in P accumulation in sorghum plant leaves. Therefore, these discrepant results show that many studies carried out under controlled conditions need to be adjusted so that they can better represent real field conditions (LACERDA, 2005; MITTLER, 2006).High concentrations of sodium ion can inhibit the absorption of K^+ by plants, through antagonism between these two ions, and potassium ion leakage can occur when Na^+ is replaced by Ca^{2+} in cell membranes (MARSCHNER, 1995). In fact, some authors have observed the existence of multiple absorption systems with different selectivities for K^+ and Na^+ , which may reflect the plant's need to coordinate the influx of these cations (SCHACHTMAN and LIU, 1999). In the present study, however, the K^+ content was not influenced by the increase in irrigation water salinity (FIGURE 10), a result similar to that obtained in sorghum by Aquino (2005), but divergent from the results obtained by most authors (IGARTUA et al., 1995; OLIVEIRA, 2002; SILVA et al., 2003). Particularly in the case of cowpea plants, conflicting results have been observed

(COSTA et al., 2003), with accumulation being observed in some genotypes and reduction in others. According to Lacerda (2005), the duration of the stress and the age of the leaf sampled can produce different results and interpretations.

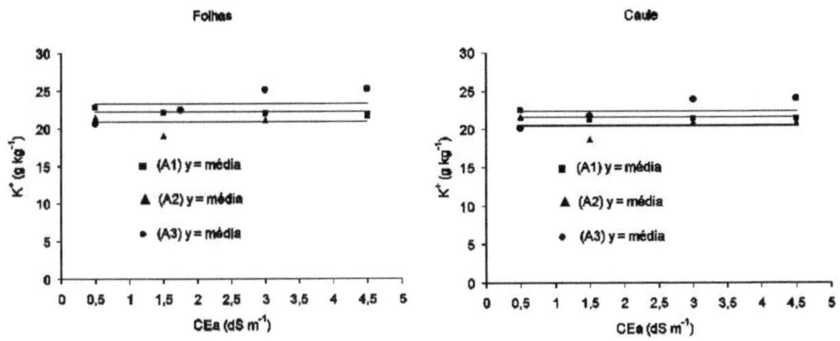

FIGURA 10 - Potassium content in the leaves and stem of string bean plants *(Vigna unguiculata* (L.) Walp.) cv. Pitiúba as a function of irrigation water salinity.

Several authors have already found an increase in potassium under salt stress conditions in leaves of maize (ERDEI and TALEISNIK, 1993), cowpea (COSTA et al., 2003; SILVA et al., 2003) and cashew (ABREU, 2004), 2003) and in cashew (ABREU, 2004); as well as reductions in the levels of this ion in leaves of maize (AZEVEDO NETO et al., 2004) and sorghum (LACERDA, 2000; NETONDO et al, 2004), and in roots of maize (AZEVEDO NETO, 1997) and citrus (RUIZ et al., 1997). On the other hand, the chemical composition of the irrigation water significantly affected the accumulation of this ion (TABLE 8). TABLE 8 shows that when Water 3 is used, the levels of potassium in the leaves and stem are higher than the levels of this ion found when Water 1 and Water 2 are used ( ). It should be noted that Water 3 is essentially composed of NaCl.

TABLE 8 - Average levels of calcium (Ca^{2+}) and potassium (K^{+}), in g kg^{-1}, in the leaves and stems of cowpea plants *(Vigna unguiculata* (L.) Walp.) cv. Pitiúba, when irrigated with water of different ionic compositions. Averages followed by the same letter in the columns do not differ by the Tukey test at 5% probability.

Treatments	Leaves		Stem	
	Ca^{2+}	K^{+}	Ca^{2+}	K^{+}
Water 1	9,63 a	22.22 ab	3,16 a	21.61 ab
Water 2	10,13a	20,82 b	2,54 b	20,52 b
Water 3	8,17 b	23,33 a	1,08 c	22,44 a

Water 1 = composed of salts of NaCl, CaCl2.2H2O and MgCl2.6H2O; Water 2 = composed of salts of Na2SO4, CaCl2, MgCl2, Na2CO3, NaHCO3 and NaCl; Water 3 = composed of salts of NaCl and CaCl2

4.4.2 Na^+ and Cl* content

The ionic composition of the irrigation water significantly affected the levels of sodium and chloride ions in both the stems and leaves (TABLE 9), while the increasing levels of salts in the water only did not influence the sodium variable in the stem, as the other variables analysed were highly influenced by salinity (TABLE 9). This was also observed by other authors (LACERDA, 1995; COSTA et al. 2003; SILVA et al., 2003; GUIMARÃES, 2005). As for the interaction between water and salt, only the variable sodium in the leaves showed statistical significance at the 1% probability level (TABLE 9).

TABLE 9 - Mean square values and statistical significance for the variables sodium (Na^+) and chlorine (Cl") in the stem and leaves of runner bean plants *(Vigna unguiculata* (L.) Walp.) cv. Pitiúba, irrigated with saline water of different ionic compositions.

Sources of Variation	Mean Square			
	Stem		Leaves	
	In^+	cr	In^+	cr
Water	0,034**	22,979**	12,389**	29,099**
Salt	0,0083**	26,816**	0.747^{ns}	39,958**
Water x Salt	0.0010^{ns}	2.674^{ns}	0,714**	1.971^{ns}
Waste	0,0053	3,338	0,041	1,495
C.V. (%)	12,94	10,41	11,46	7,91

The increase in sodium content in the stem when salt levels were increased is shown in FIGURE 11. It is worth pointing out that the sodium content in the stem was higher than in the leaves, a result also found by Lacerda (1995), Silva (1998) and Guimarães (2005), also in Pitiúba cowpea plants. This can be explained by the fact that during its transport through the plant, i.e. from the roots to the leaves, this ion is eliminated from the transpiratory flow to be compartmentalised in the stem + petiole tissues (SILVA, 1998). In addition, the sodium ion can be readily translocated through the phloem, so the high concentrations that occur in actively transpiring leaves can be diluted through redistribution throughout the plant (LARCHER, 2000). Generally, the lower accumulation of sodium in leaves is associated with the relative ability of glycophytes to limit the absorption and/or transport of toxic ions, mainly Na^+ and Cl", from the root zone to the aerial part, preventing their accumulation at levels that exceed the cells' ability to compartmentalise them in the vacuole (GREENWAY and MUNNS, 1980; MUNNS, 2002).

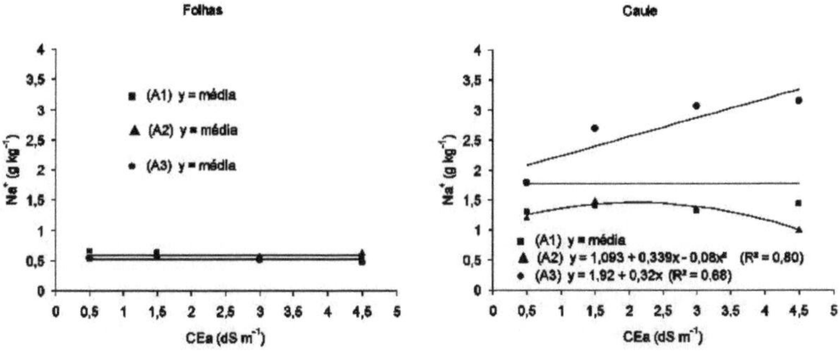

FIGURA 11 - Sodium content in the stem and leaves of string bean plants *(Vigna unguiculata* (L.) Walp.) cv. Pitiúba as a function of irrigation water salinity.

When analysing the response of the chloride ion to salinity, we can see that there was an increase in its levels in relation to the treatment with the lowest level of salts (FIGURE 12), and compared to the sodium ion, they were higher, most likely due to the chemical composition of the irrigation water used in the experiment. A similar result was found by Guimarães (2005) in string bean plants of the same cultivar used in this experiment, observing that, for any organ analysed, the chloride content was always higher than the sodium content. The distribution of chlorine in the plant was more uniform than sodium (FIGURE 12). According to Larcher (2000), the redistribution of the chloride ion by the plant is a way of surviving in saline soils. The accumulation of chloride ion in the leaves can contribute to osmotic adjustment (BERNSTEIN, 1961, 1963; SLATYER, 1961, SILVA, 1998) and increase leaf juiciness as a result of the accumulation of water inside the cells and intercellular spaces (STROGONOV, 1964; SLATYER, 1967; MERCADO, 1973; ZEKRI and PARSONS, 1990; NOBLE and ROGERS, 1994; SILVA, 1998), but it can slow down or inhibit the formation of new cells (SILVA, 1998).

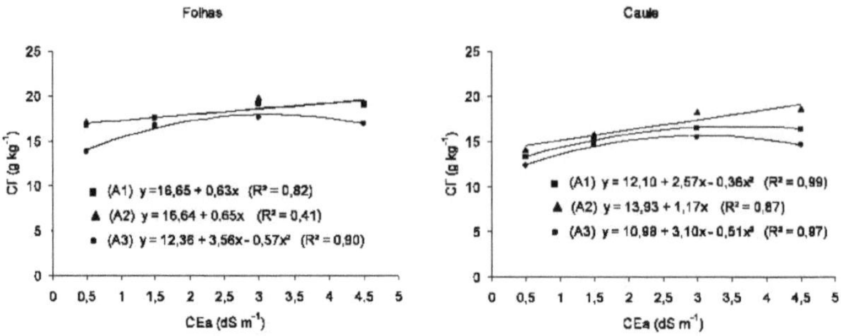

FIGURA 12 - Chlorine content in the leaves and stem of string bean plants *(Vigna unguiculata* (L.) Walp.) cv. Pitiúba as a function of irrigation water salinity.

TABLE 10 shows the influence of the chemical composition of the irrigation water on the accumulation of sodium in both the leaves and the stems. It can be seen that Water 3 has higher sodium levels in the stems than Waters 1 and 2, which is not the case for the accumulation of this ion in the leaves, as Water 1 and Water 2 did not differ statistically, showing higher values than Water 3. This result can be explained by the fact that the chemical composition of water 3 is only NaCl; indicating that for there to be efficient compartmentalisation of this ion, there must be a predominance in relation to the other ions in the composition of the water used for irrigation; since in the composition of waters 1 and 2, used in this experiment, there are salts of calcium, magnesium as well as the presence of sulphates, carbonates and bicarbonates. It can also be seen in TABLE 10 that the chemical composition of waters 2 and 1 showed higher values for chlorine ion in the leaves, while water 3 showed a lower value for this ion. In the stem, the values were higher for water 2, followed by water 1 and, lastly, water 3.

TABLE 10 - Average levels of sodium (Na^+) and chlorine (Cl), in g kg^{-1}, in the leaves and stems of string bean plants *(Vigna unguiculata* (L.) Walp.) cv. Pitiúba, when irrigated with water of different ionic compositions. Averages followed by the same letter in the columns do not differ by the Tukey test at 5% probability.

Treatments -	Leaves		Stem	
	In^+	Cl^-	In^+	Cl^-
Water 1	0,60 a	18,16a	1,36 b	15,37 b

Water 2	0,58 a	18,19a	1,26 b	16,70 a
Water 3	0,52 b	16,32 b	2,67 a	14,29 c

Water 1 = composed of salts of NaCl, CaCl2.2H2O and MgCl2.6H2O; Water 2 = composed of salts of Na2SO4, CaCl2, MgCl2, Na2CO3, NaHCO3 and NaCl; Water 3 = composed of salts of NaCl and CaCl2

4.4.3 MICRONUTRIENT CONTENT

Analysing the absorption of micronutrients (Cu, Fe, Mn and Zn) in Pitiúba cowpea plants when subjected to saline stress with water of different ionic compositions, we can see from TABLE 11 that the chemical composition of the irrigation water influenced the absorption of copper, iron and manganese ions, but did not influence zinc ion concentrations, both in the stem and in the leaves.

TABLE 11 - Mean square values and statistical significance for the variables copper (Cu), iron (Fe), manganese (Mn) and zinc (Zn) in the stem and leaves of cowpea plants *(Vigna unguiculata* (L.) Walp.) cv. Pitiúba, irrigated with saline water of different ionic compositions.

Sources of Variation	Mean Square[1]			
	Stem			
	Ass	Fe	Mn	Zn
Water	0,374035**	0,051023**	1,585664**	0.251795^{ns}
Salt	0.041968^{ns}	0.001443^{ns}	0,304197**	0.236532^{ns}
Water x Salt	0.066999^{ns}	0,020536**	0,113118**	0.216050^{ns}
Waste	0,031877	0,005536	0,014380	0,337005
C.V. (%)	10,64	4,25	7,69	19,01
	Leaves			
Water	0,405917**	0,068937**	0,456245**	0.109038^{ns}
Salt	0.089826^{ns}	0,074428**	0,124068**	0.038884^{ns}
Water x Salt	0,193727*	0,101151**	0,025815**	0.254235^{ns}
Waste	0,069972	0,012636	0,006794	0,233673
C.V. (%)	13,34	5,45	3,72	15,08

'Significant by F-test at 5%; ** Significant by F-test at 1%; ns = not significant.
[1] Values transformed into Neperian Logarithm.

Increasing salinity only influenced the iron (stem) and manganese (leaf and stem) ions at the 1% probability level (TABLE 11). The interaction between the ionic composition of the water and salinity showed different results, since in the stem, the iron and manganese ions were statistically significant at the 1% probability level, while in the leaves (TABLE 11), the copper, iron and manganese ions were statistically significant at the 1% probability level (TABLE 11).

TABLE 12 shows that the copper ion content in the leaves was higher when irrigation water with calcium, magnesium and sodium salts was used (A1), followed by water composed of CaCh and bicarbonates (A2) and water with only NaCl in its composition (A3); on the other hand, in the stem, the highest copper ion contents were found in Water 1, followed by Water 3 and 2. The iron ion levels in the leaves were higher in plants irrigated with Water 3, followed by Waters 1 and 2. In the stem, higher values were observed when irrigating with Waters 1 and 3 compared to Water 2. When analysing the average manganese ion levels in the leaves and stem of the cowpea plants (TABLE 12), a similar behaviour was observed to that found for Fe in the stem, i.e. water 1 and 3 showed higher values than water 2. This result suggests that the acquisition of micronutrients such as iron and manganese by the plant can be affected by the composition of the saline environment, and a reduction in the absorption of these micronutrients can be observed due to antagonism with excess calcium ion. A similar result was found by Alvarenga et al. (2000), who observed a reduction in manganese absorption in lettuce when calcium was applied foliarly.

TABLE 12 - Average contents[1] of copper (Cu), iron (Fe) and manganese (Mn), in mg kg^{-1}, in the leaves and stems of cowpea plants *(Vigna unguiculata* (L.) Walp.) cv. Pitiúba, when irrigated with water of different ionic compositions.

	Leaves			Stem		
Treatments	Ass	Fe	Mn	Ass	Fe	Mn
Water 1	2,144 a	2,006 a	2,317 a	1,834 a	1,783 a	1,743 a
Water 2	1,875 b	2,053 b	2,044 b	1,580 b	1,692 b	1,235 b
Water 3	1,875 b	2,122 ab	2,292 a	1,620 b	1,775 a	1,701 a

Water 1 = composed of NaCl, CaCl2.2H2O and MgCl2.6H2O salts; Water 2 = composed of Na2SO4, CaCl2, MgCl2, Na2CO3, NaHCO3 and NaCl salts; Water 3 = composed of NaCl and CaCl2 salts * Means followed by the same letter in the columns do not differ by the Tukey test at 5% probability.
[1] Values transformed into Neperian Logarithm.

In FIGURE 13, we can see that the increase in salinity caused a decrease in manganese levels, both in the leaves and in the stem, with a quadratic behaviour for water 1 (A1) and water 2 (A2) and a linear behaviour for water 3 (A3). Also, in the same figure, we see that the iron ion does not suffer a marked reduction as the salt levels increase, in waters 2 and 3, while in water 1, there is an increase in the concentration of this micronutrient.

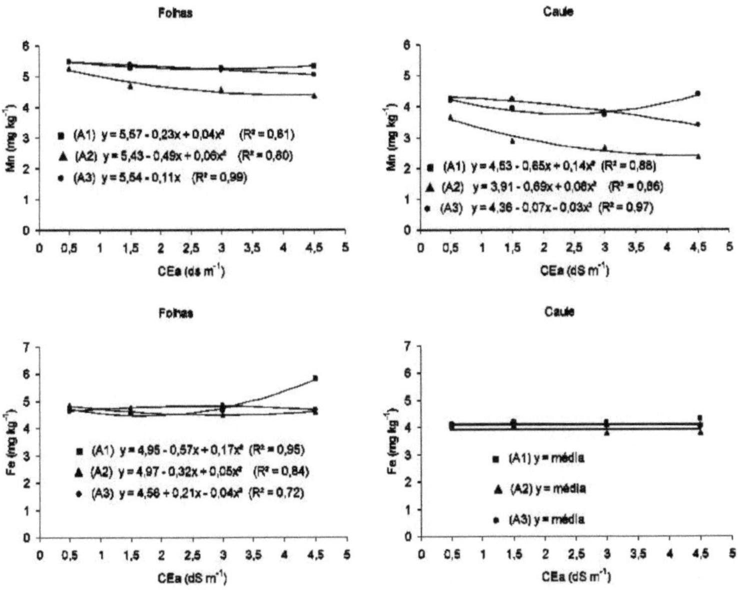

FIGURA 13 - Iron and manganese content in leaves and stems of cowpea plants *(Vigna unguiculata* (L.) Walp.) cv. Pitiúba as a function of irrigation water salinity. Values transformed into Neperian logarithms.

In general, the lowest micronutrient levels were found in plants irrigated with Water 2, which is rich in bicarbonates and calcium salts. This can be explained, in part, by the fact that micronutrient cations (Fe, Cu. Mn and Zn) are greatly influenced by the pH and calcium content of the soil, Mn and Zn) are greatly influenced by the pH and calcium content of the soil, i.e. in acidic conditions, micronutrient cations are more soluble and available for absorption, and can sometimes become toxic to plants. As the pH rises, these ions change to insoluble forms (oxides and hydroxides), reducing their solubility and availability (AQUINO, 2003).

4.5 Accumulation of organic solutes in the leaves

In TABLE 13, we can see that salinity only significantly influenced the carbohydrate content of Pitiúba cowpea leaves, since the variations in proline, protein and N-aminosoluble content were not statistically significant. For the composition of the irrigation water, the carbohydrate and protein contents were statistically significant at 1% and 5% probability, respectively, while the proline and N-aminosoluble values were not statistically significant (TABLE 13). The interaction between water and salinity did not statistically influence any of the variables evaluated (TABLE 13).

TABLE 13 - Mean square values and statistical significance for the variables proline (PROLINE), carbohydrates (CHOS), proteins (PROT) and N-aminosoluble (AMINO) in the leaves of cowpea plants irrigated with saline water of different ionic compositions.

Sources of Variation	Mean Square			
	PROLINA	CHOS	PROT	AMINO
Water	5.295^{ns}	4000,082**	1,055*	1987.25^{ns}
Salt	2.045^{ns}	2118,668*	0.754^{ns}	948.88^{ns}
Water x Salt	2.754^{ns}	608,530^{ns}	0.481^{ns}	1485.51^{ns}
Waste	1,734	625,183	0,301	1503,58
C.V. (%)	82,43	5,35	31,46	54,39

'Significant by F-test at 5%; ** Significant by F-test at 1%; ns = not significant

Soluble carbohydrates and amino acids help to protect biomembranes and proteins from the deleterious effects of high ionic concentration (LARCHER, 2000). FIGURE 14 shows the influence of increasing salt levels on the carbohydrate content of cowpea leaves. It can be seen that at first there was a small reduction followed by an increase, but without there being a greater discrepancy between the highest salinity level (4.5 dS $m^{-1)}$ and the lowest salinity level (0.5 dS $m^{-1)}$. A similar result was found by Costa et al. (2003), when analysing the concentration of organic solutes in various cultivars of string beans, including the Pitiúba cultivar. This result indicates that the increase in carbohydrate content did not seem to have contributed significantly to maintaining a water potential favourable to water absorption, i.e. leaf osmotic adjustment.

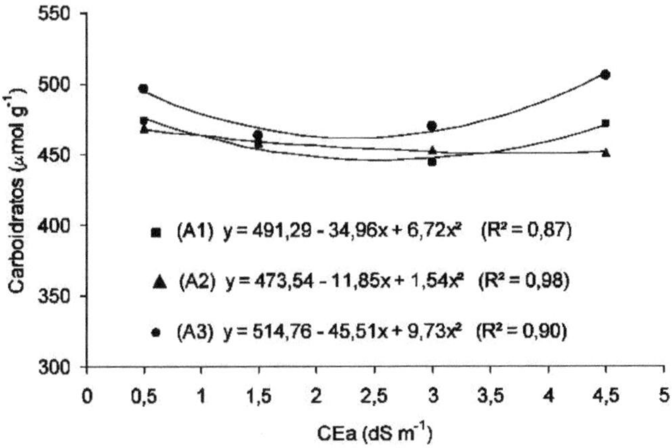

FIGURA 14 - Carbohydrate content in the leaves of cowpea plants *(Vigna unguiculata* (L.) Walp.) cv. Pitiúba as a function of irrigation water salinity.

TABLE 14 shows that the composition of Water 3, which is rich in sodium chloride, had an influence on obtaining the highest carbohydrate values, surpassing Waters 1 and 2, which in turn were statistically equal. Silva (1998), when using NaCl on string bean plants, also observed an increase in carbohydrates in the leaves. For the average protein values, we can see that the composition of Water 1 obtained the highest values, followed by the composition of Water 3 and, lastly, Water 2 (TABLE 14).

TABLE 14 - Average carbohydrate (CHOS), N-aminosoluble (AMINO), proline (PROLINE) and protein (PROT) contents in leaves of cowpea plants *(Vigna unguiculata* (L.) Walp.) cv. Pitiúba, when irrigated with water of different ionic compositions.

Treatments	CHOS	AMINO	PROLINA	PROT
	(µmol g^{-1})			(mg g^{-1})
Water 1	461,63 b	74,34 a	1,62 a	1,96 a
Water 2	457,64 b	60,16 a	1,07 a	1,50 b
Water 3	483,89 a	79,38 a	2,10a	1.77 ab

Water 1 = composed of NaCl, CaCl2.2H2O and MgCl2.6H2O salts; Water 2 = composed of Na2SO4, CaCl2, MgCl2, Na2CO3, NaHCO3 and NaCl salts; Water 3 = composed of NaCl and CaCl2 salts Averages followed by the same letter in the columns do not differ by the Tukey test at 5% probability.

In the process of osmotic adjustment, plants absorb ions and accumulate various organic solutes in their cells, resulting in a lowering of the osmotic potential and maintenance of the water potential gradient between the external environment and the inside of the plant (TAIZ and ZEIGER, 2002). In glycophytes, osmotic adjustment also occurs, but the compartmentalisation of ions is not as efficient as in halophytes (O'LEARY, 1975). According to Greenway and Munns (1980), glycophytes generally retain large quantities of salts in their stems and roots, so osmotic adjustment in the leaves depends more on the accumulation of organic solutes. Among the organic regulators found in higher plants are: alcohols, soluble sugars, betaines, amino acids, proline and derivatives (SOMERO, 1992; HANSON, 1992; SERRANO, 1996; GARCIA et al., 1997; ZHU et al., 1997; IYER and CAPLAN, 1998).

CHAPTER 5

CONCLUSIONS

- Salinity significantly reduced biomass production, leaf growth and gas exchange in the plants, although the composition of the water had little influence on these parameters;
- Salinity caused a greater accumulation of Ca, P and Cl ions in the leaves; however, in the stem there was a decrease in the concentration of Ca^{2+}, an increase in Na^+ , Cl^- and P.
- The effect of the ionic composition of the irrigation water was observed mainly on mineral nutrition;
- The absorption of micronutrients (Cu, Fe, Mn and Zn) was affected by the greater presence of calcium salts in the irrigation water as well as by the increase in pH.
- The accumulation of potassium was not influenced by the increasing salinity of the irrigation water, but the ionic composition did influence the concentration of this ion in the plant;
- No significant relationships were observed between the levels of organic solutes analysed and the salinity levels and chemical composition of the water.
- The accumulation of inorganic solutes, such as Na^+ , Ca^{2+} and Cl^-, suggests that they are involved in the plant's osmotic adjustment;
- The chemical composition of irrigation water is important when you want to irrigate, as the addition of different salts to the soil can have a specific influence on the absorption of mineral nutrients, with negative repercussions on the plant's metabolism and productivity.

REFERENCES

ABREU, C.E.B. **Growth, osmoregulation and mobilisation of reserves in early dwarf cashew seedlings subjected to salinity.** Federal University of Ceará, 2004. 60p. Monograph (Graduation in Agronomy)

ALBERICO, G. J.; CRAMER, G. R.. Is the salt tolerance of maize related to sodium exclusion.I.Preliminary screening of seven cultivars. J. of Plant Nutr.,v.16, p. 2289-2303, 1993.

ALVARENGA, M.A.R.; SILVA, E.C.; SOUZA, R.J.;CARVALHO, J.G. Effect of nitrogen doses applied to the soil and calcium levels applied via foliar application on the content and accumulation of micronutrients in American lettuce. **Ciênc. agrotec.,** Lavras, v.24, n.4, p.905-916, Oct./Dec., 2000.

AMORIM, S.M.C; VALLE, R.R. Absorption and resistance to water movement in cocoa trees. **Pesq. Agropec.** Brasília**,** v.28, n.8, p.907-913, august,1993.

ANDRADE, S.M.G. **Effect of Ca^{2+} on germination, mineral composition and growth of rice *(Oryza sativa* L.) grown in NaCl solution.** Federal University of Ceará, Fortaleza. 94p. 1989. Master's dissertation in Soils and Plant Nutrition.

AQUINO, B.F. **Fundamental concepts in soil fertility.** Fortaleza: UFC, 2003. 182p. Class notes

AQUINO, A.J.S. **Evaluation of growth and salinity tolerance mechanisms in forage sorghum plants irrigated with saline water.** Fortaleza: UFC, 2005, 73p. Master's dissertation

ARAGÃO, M.E.F.; GUEDES, M.M.; OTOCH, M.L.O.; GUEDES, M.I.F.; MELO, D.F.; LIMA, M.G.S. Differential responses of ribulose-1,5- bisphosphate carboxylase/oxygenase activities of two *Vigna unguiculata* cultivars salt stress. **Braz. Journal Plant Physiology,** v.17, n.2, p.207-212, 2005.

ARAGÚES, R. Irrigated agriculture, water quality and return flows. In: **Symposium Nacional Presente y Futuro de los Regadios Espanoles.** CEDEX- MOPTMA. Madrid, Jun/1994. 18p.

ARAGÚES, R. Agricultura de regadio y salinización de suelos y aguas. **Frontiers of Science and Technology,** n.8:,p.36-39, 1995.

ARAÚJO, J.P.P.; WATT, E.E. **Caupi in Brazil.** EMBRAPA/CNPAF. Brasília, DF, 1988. 722 p.

AYERS, R. S.; WESTCOT, D. W. **Water quality in agriculture.** Trad. H. R. Gheyi et al., Campina grande: UFPB, 1999. 153p. (FAO Studies: Irrigation and Drainage, 29)
AZAIZEH, H.; GUNSE, B.; STEUDLE, E. Effects of NaCl and $CaCl_2$ on water transport across root cell of maize (Zea *mays* L.) seedlings. **Plant Physiol.,** v.99, p.886-894, 1992.

AZEVEDO NETO, A.D. **Study of growth and nutrient distribution in maize seedlings subjected to salt stress.** UFRPE, Recife. 133p. 1997. Master's dissertation.

AZEVEDO NETO, A.D.; PRISCO, J.T.; ENÉAS FILHO, J.; LACERDA, C.F.; SILVA, J.V.; COSTA, P.H.A.; GOMES FILHO, E. Effects of salt stress on plant growth, stomatal response and solute accumulation of different maize genotypes. **Brazilian Journal of Plant Physiology,** v. 16, p. 31-38, 2004.

BATES, L. S.; WALDREN, R. P.; TEARE, I. D. Rapid determination of free proline for water-stress studies. **Plant and Soil,** v. 39, p. 205-207, 1973.

BERNARDO, S. Environmental impact of irrigation in Brazil. In: SILVA, D. D.; PRUSKI.F.F. (Eds.). **Water Resources and the Sustainable Development of Agriculture.** Viçosa: MMA, SRH, ABEAS, UFV, 1997, 252p.

BERNARDO, S.; SOARES, A. A.; MANTOVANI, E. C. **Manual de irrigação.** 7 ed. Viçosa: Ed. UFV. 2005. 611p.

BERNSTEIN, L. Osmotic adjustment of plants to saline media. I. Steady State. **Amer. J. Bot.,** v.48, p.909-918, 1961.
BERNSTEIN, L. Osmotic adjustment of plants to saline media. II. Dynamic Phase. **Amer. J. Bot.,** v.50, p.360-370, 1963.

BIGGAR, J.W.; ROLSTON, D.E.; NIELSOEN, D.R. Transport of salts by water. **California Agriculture,** v.38, n.10, p.10-11. 1984.

BLANCO, F. F. **Tolerance of grafted cucumbers to salinity in a protected environment and control of soil salinisation.** ESALQ: Piracicaba - SP. 1999. 104p (Master's dissertation)

BOHN, H.L.; McNEAL, B.L.; O'CONNOR, G.A. **Soil Chemistry.** New York, John Willey & Sons, 1985. 341 p.

BRADFORD, M. M. A rapid and sensitive method for the quantification of microgram quantities of protein utilizing the principie of protein-dye binding. **Analytical Biochemistry,** v.72, p.246-254, 1976.

BUCHANAM, B.B.; GRUÍSSEM, W.; JONES, R.L. **Biochemistry and molecular biology of plants.** Rockville, Maryland: American Society of Plant Physiologists, 2000. 1367p.

CACHORRO, P.; ORTIZ, A.; CERDÁ, A. Effects of saline stress and calcium on lip composition in bean roots. **Phytochem.,** v.32,p.1131-1136, 1993.

CACHORRO, P.; ORTIZ, A.; CERDÁ, A. Implications of calcium nutrition on the response of *Phaseolus vulgaris* L. to salinity. **Plant and Soil,** v.159, p.205-212, 1994.

CALEGARI, A. Legumes for summer green manure in Paraná. Londrina: IAPAR,

1995. p.115. Technical Circular, 80.

CARMO, G. A; MEDEIROS, J. F.; TAVARES, J. C.; GHEYI, H. R.; SOUZA, A. M.; PALÁCIO, E. A. Q. Growth of banana trees under different levels of irrigation water salinity. **Rev. Bras. Frutic.,** Jaboticabal/SP, v.25, n.3, p.513-518, 2003.

CARVALHO, J.A.; PEREIRA, G.M.; ANDRADE, M.J.B.; ROQUE, M.W. Effect of water deficit on the yield of cowpea *(Vigna unguiculata* (L.) Walp). **Ciênc. agrotecnica**, Lavras, v.24, n.3, p.710-717, jul./set., 2000.

CHAUNDARY, M. T.; WAINWRIGHT, S. J.; MERRETT, M. J.; SHAHEALAM, M. Salt tolerant plants of Lucerne *(Medicago media* Pers) regenerated from salt-selected suspension cultures. **Plant Science,** v.98, p.97-102, 1994.

CHEESEMAN, J.M. Mechanisms of salinity tolerance in plants. **Plant Physiology,** v.87, p.547-550, 1988.

COLMER, T. D.; FAN, T. W. M.; HIGASHI, R. M.; LAUCHLI, A.. Interactions of Ca^{2+} and NaCl stress on the ion relations and intracellular pH of *Sorghum bicolor* root tips: an in vivo P-31-NMR study. **J. Exp.Bot.,** v.45, p.1037-1044, 1994.

COLMER, T.D.; FAN, T.W.M., HIGASHI, R.M.; LÃUCHLI, A. Interactive effects of Ca^{2+} and NaCl salinity on the ionic relations and proline accumulation in the primary root tip of *Sorghum bicolor.* **Physiol. Plant,** v.97, p.421-424, 1996.

CONTRERAS, A.M.; ELIZONDO, M.S. **Relaciones agua-suelo-planta- atmósfera.** Mexico: Universidad Autonoma de Chapingo, Departamento de Ensenanza Investigación y Servicio en Irrigación, 2. ed.,1980, 22p.

COSTA, P.H.A.; SILVA, J.V.; BEZERRA, M.A.; ENÉAS FILHO, J.; PRISCO, J.T.;

GOMES FILHO, E. Growth and levels of organic and inorganic solutes in cultivars of *Vigna unguiculata* submitted to salinity. **Revista Brasil. Bot.,** v.26, n.3, p.289-297, jul-set., 2003.

CRAMER, G.R.; EPSTEIN, E.; LÃUCHLI, A. Na-Ca interactions in barley seedlings: relationship to ion transport and growth. **Plant, Cell and Environ.,** v.12, p.551-558, 1989.

CRAMER, G.R.; EPSTEIN, E.; LÃUCHLI, A. Effects of sodium, potassium and calcium on salts-stressed barley. I Growth analysis. **Physiol. Plant,** v.80, p.83-88, 1990.

CRAMER, G.R.; EPSTEIN, E.; LÃUCHLI, A. Effects of sodium, potassium and calcium on salt stressed barley. IL- Elemental analysis. **Physiol. Plant,** v.81, p. 197-202, 1991.

CRUCIANI, D.E. Agricultural drainage for the recovery of soils affected by salts. In: GHEYI, H.R.; QUEIROZ. J.E.; MEDEIROS, J.F. (Eds.) **Management and control of salinity in irrigated agriculture.** Campina Grande: UFPB- SBEA, 1997. Chap.8, p.239-284.

CRUZ, W.B.; MELO, F.A.C.F. Chemical zoning and salinisation of groundwater in the Northeast of Brazil. In: SUDENE, **Boletim de Recursos naturais,** v.7, n. %, p.7-40, jan/dez, 1969.

DANTAS, J.P.; MARINHO, F. J. L.; FERREIRA, M.M.M.; AMORIM, M.S.N.; ANDRADE, S.LO.; SALES, A.L. Evaluation of caupi genotypes under salinity. **Revista Brasileira de Engenharia Agrícola e Ambiental,** Campina Grande, v.6, n.3, p.425-430, 2002.

DIAS, N. S. **Fertigation management and salinity control in soil cultivated with**

lacy melon under protected environment. Piracicaba: ESALQ/USP. 2004. 110p. (Doctoral Thesis)

DUBOIS, M.; GILLES, K. A.; HAMILTON, J. K.; REBERS, P. A.; SMITH, F. Colorimetric method for determination of sugars and related substances. **Analytical Chemistry,** v.28, p.350-356, 1956.

BRAZILIAN AGRICULTURAL RESEARCH COMPANY. **Cultivation of cowpeas**, Teresina, 2003. Embrapa Meio Norte. Production systems, 2. Electronic version

EPSTEIN, E. **Nutrição mineral de plantas: princípios e perspectivas.** São Paulo, EDUSP, 341 p. 1975.
ERDEI, L.; TALEISNIK, E. Changes in water relation parameters under osmotic and salt stresses in maize and sorghum. **Physiologia Plantarum,** v.89, p.381-387, 1993.

FAGERIA, N.K. **Tropical soils and physiological aspects of crops.** Brasília: EMBRAPA/DPU, 1989. 425p. (EMBRAPA - CNPAF. Documentos, 18)

FAGERIA, N.K.; GHEYI, H.R. Genetic improvement of crops and selection of cultivars. In: GHEYI, H.R.; QUEIROZ. J.E.; MEDEIROS, J.F. (Eds.) **Management and control of salinity in irrigated agriculture.** Campina Grande: UFPB-SBEA, 1997. Chap. 11, p. 363-383.

FAO. Faostat agriculture. 2004. Available at **http://faostat.**fao.org/faostat/.(07 November 2005)

FNP CONSULTORIA & COMERCIO. **Agrianual 2004:** yearbook of Brazilian agriculture. São Paulo. 2004. 546 p.

GAINES, T. P.; PARKER, M. B.; GASCHO, G. J. Automated determination of chlorides in soil and plant tissue by sodium nitrate. **Agronomy Journal,** v.76, p.371-374, 1984.

GALVEZ, A. F.; GULICK, P. J.; DVORAK, J. Characterisation of the early stages of genetic salt-stress responses in salt-tolerant *Lophopyrum elongatum*, salt-sensitive wheat, and their amphiploid. **Plant Physiol.,** v.103, p.257-265, 1993.

GARCIA, A.B.; ENGLER, J.A.; IYER, S.; GERATS, T.; MONTAGU, M.V.; CAPLAN, A.B. Effects of osmoprotectants upon NaCl stress in rice. **Plant Physiol.,** v.115, p. 159-169, 1997.

GHEYI, H.R.; MEDEIROS, J.F.; BATISTA, M.A.F. Prevention, management and recovery of saline and sodic soils. **Specialisation Course in Irrigation and Drainage.** Mossoró: ESAM, 1991. 56p. Handout

GRATTAN, S.R.; MAAS, E.V. Interactive effects of salinity and substrate phosphate on soybean. **Agron. J.,** v.76, p.668-676, 1984.

GREENWAY, H.; MUNNS, R. Mechanisms of salt tolerance in nonhalophytes. **Ann. Rev. Plant Physiol.,** v.31, p.149-190, 1980.

GRIEVE, C.M.; FUJIYAMA, H. The response of two rice cultivars to externai Na/Ca ratio. **Plant and Soil,** v.103, p.245-250, 1987.

GRIEVE, C.M.; MAAS, E.V. Differential effects of sodium/calcium ratio on sorghum genotypes. **Crop. Sei.,** v.28, p.659-665, 1988.

GUILHERME, E.A., LACERDA, C.F., BEZERRA, M.A., GUIMARAES, F.V.A., PRISCO, J.T., ENÉAS FILHO, J" GOMES FILHO, E. Effects of salt stress on leaf growth and carbon assimilation rate in cowpea plants. In: IX Brazilian Congress of Plant Physiology, 2003, Atibaia. **Brazilian Journal of Plant Physiology**

(Supplement). Londrina: Brazilian Society of Plant Physiology, v.15, p.391,2003.

GUIMARÃES, F.V. **Physiological and biochemical responses in string bean plants cv. Pitiúba submitted to NaCl stress at different concentrations of CaC_2 and CaSCh.** Fortaleza - UFC, 2005. 176p. PhD Thesis in Plant Biochemistry.

HANSON, A.D. Compatible solute synthesis and compartmentalisation in higher plants. In: **Water and Life.** SOMERO, G.N.; OSMOND C.B.; BOLIS C.L. (eds). Springer-Verlag, Berlin-Heidelberg. Germany, 1992. p.52-69.

HOLANDA, J. P.; AMORIM, J.R.A. Water quality for irrigation. In: GHEYI, H.R.; QUEIROZ, J.E.; MEDEIROS, J.M. (eds) **Manejo e Controle da salinidade na agricultura irrigada.** Campina Grande: UFPB-SBEA, 1997. Chap.5, p.137-169.

HOORN, J. W. van. Quality of irrigation water, limits of use of long-term effects. In: **Salinity Seminar, Baghidad.** Rome: FAO, 1971, p.117-35. (FAO: Irrigation and Drainage Paper, 7)

HUANG, J.; REDMANN, R.E. Solute adjustment and calcium supply in cultivated and wild barley. J. **Plant Nutrition,** v.18, n.7, p. 1371-1389, 1995.

IGARTUA, E.; GRACIA, M. P.; LASA, J. M. Field responses of grain sorghum to a salinity gradient. **Field Crop Research,** v.42, p. 15-25, 1995.

IYER, S.; CAPLAN, A. Products of proline catabolism can induce osmotically regulated genes in rice. **Plant Physiol.,** v.116, p.203-211, 1998.

KENT, L.M.; LÃUCHLI, A. Germination and seedling growth of cotton: salinity-calcium interactions. **Plant Cell Environm.,** v.8, p. 155-159, 1985.

KOVDA, V. A.; YARON, B.; SHALHEVELT, Y. Quality of irrigation water. In: KOVDA,

V. A.; BERG, C. van den, HAGAN, R. M. (Eds.). **Irrigation, drainage and salinity.** London: HUTCHINSON/FAO/UNESCO, 1973. Chap. 7, p. 177-205. (FAO-UNESCO. International Source Book)

LACERDA, C.F. **Effects of salinity on the development and mineral composition of string beans *(Vigna unguiculata* (L.) Walp.) and the use of Ca^{2+} as a means of minimising these effects.** Fortaleza : UFC, 1995, 87p. (Master's dissertation in Soils and Plant Nutrition)

LACERDA, C.F. **Growth and accumulation of organic and inorganic solutes in two forage sorghum genotypes subjected to salt stress.** Viçosa:UFV, 2000. 163p. (Doctoral Thesis in Plant Physiology)

LACERDA, C.F.; CAMBRAIA, J.; CANO, A.O.; RUIZ, H.A.; PRISCO, J.T. Solute accumulation and distribution during shoot and leaf development in two sorghum genotypes under salt stress. **Environmental and Experimental Botany,** v.49, p. 107-120, 2003.

LACERDA, C.F. Salinity x mineral nutrition interaction. In: NOGUEIRA, R.J.M.C., ARAÚJO, E.L., WILLADINO, L.G., CAVALCANTE, U.M.T. (eds.) **"Environmental stresses: damage and benefits in plants".** Recife, UFRPE, p. 127-137, 2005.

LAHAYE, P.A.; EPSTEIN, E. Calcium and salt tolerance by bean plants. **Physiol. Plant,** v.25, p.213-218, 1971.

LARCHER, W. **Plant Ecophysiology.** Translation: PRADO, C.H.B.A. & FRANCO. A.C. São Carlos, RiMa, 2000. 533p.

LEIDI, E.O.; NOGALES, R.; LIPS, S.H. Effect of salinity on cotton plants growth under nitrate or ammonium nutrition at different calcium leveis. **Field Crop Res.,** v.26, p.35-44, 1991.

LEMOS, G.B. **Evaluation of the nutritional status of two varieties of caupi *(Vigna unguiculata* (L.) Walp) submitted to different levels of salinity.** Areia: UFPB, 1991. 85p. Undergraduate Monograph

LEOPOLD, A.C.; WILLING, R.P. Evidence for toxicity effects of salt membranes. In : **"Salinity tolerance in plants. Strategies for crop improvement".** STAPLES, R.C.; TOENNIESSEN, G.H. (eds.), John Wiley & Sons. 1984.

LEPRUN, J.C. **First evaluation of surface waters in the Northeast.** In: SUDENE. Report of the agreement on soil management and conservation in the Brazilian Northeast. Recife, 1983. p.91-141.

LEVITT, J. **Responses of plants to environmental stresses. Water, radiation, salt, and other stresses.** Vol II. 2nd edition. Academic Press, New York, 1980.

LIN, C. C.; KAO, C. H. NaCl stress in rice seedlings - The influence of calcium on root growth. **Botanical Bulletin of Academia Sinica,** v.36, p.41- 45, 1995.

LOPEZ, P.I. Groundwater quality and regulations for different uses: contamination and protection. In: **V MÁSTER INTERNACIONAL DE RIEGO Y DRENAJE.** CENTER-MAPYA, Madrid. 67p. 2000. Handout

MAAS, E.V. Salt tolerance of plants. **Applied Agricultural Research,** v.1, p. 12-26, 1986.

MAAS, E.V.; GRIEVE, C.M. Sodium-induced deficiency in salt-stressed corn. **Plant, Cell Environm.**, v.10, p.559-564, 1987.

MAAS, E.V.; HOFFMAN, G.J. Crop salt tolerance: current assessment. **Journal of**

the Irrigation and Drainage Division, A.S.C.E., v.103, n.IR2, p. 115-134, 1977

MAAS, E.V.; POSS, J.A.; HOFFMAN, G.J. Salinity sensitivity of sorghum at three growth stages. **Irrig. Sei.,** v.7, p.1-11,1986.

MALAVOLTA, E.; VITTI, G.C.; OLIVEIRA, S.A. Avaliação do estado nutricional das plantas: Princípios e Aplicações. **Brazilian Association for Potash and Phosphate Research.** Piracicaba, SP, 1989. 201 p.

MANTOVANI, A. A method to improve leaf succulence quantification. **Brazilian Archives of Biology and Technology,** v.42, p.9-14, 1999.

MARSCHNER, H. **Mineral nutrition of higher plants.** 2 Ed. Academic Press, London, 1995.

Mc DERMIT, D.K. Sources of error in the estimation of stomatal conductance and transpiration from porometer data. **HortScience,** Alexandria, v.25, n.12, p. 1538-48, 1990.

MEDEIROS, J. F.; GHEYI, H. R. **The quality of irrigation water.** Campina Grande: UFPB, 1994. 60p.

MEDEIROS, J.F.; GHEYI, H.R. Management of a soil-water-plant system in soils affected by salts. In: GHEYI, H.R.; QUEIROZ. J.E.; MEDEIROS, J.F. (Eds.) **Management and control of salinity in irrigated agriculture.** Campina Grande: UFPB-SBEA, 1997. chap.8, p.239-287.

MERCADO, A. **Structure and function of plan cells in saline habitats.** Halsted Press. Jerusalem, Israel. 284p. 1973.

MITTLER, R. Abiotic stress, the field environment and stress combination. **Trend in**

Plant Science, v.11, p. 15-19, 2006.

MIYAZAWA, M.; PAVAN, M. A. ; BLOCH, M. F. M. Evaluation of methods with and without digestion for extraction of elements in plant tissues. **Ciência e Cultura,** v.36, p. 1953-1958, 1984.

MOLLE, F.; CADIER, E. **Manual do pequeno açude.** Recife: SUDENE/DPG/PRN/DPP/APR, 1992. 523p.
MOUSINHO, F.E.P. **Economic viability of caupi beans in the state of** Piauí. ESALQ, Piracicaba - SP, 103p. 2005. Doctoral thesis.

MUHAMMED, S.; AKBAR, M.; NEUE, H.U. Effect of Na/Ca and Na/ K ratios in saline culture solution on the growth and mineral nutrition of rice *(Oryza sativa* L.). **Plant and Soil,** v.104, p.57-62, 1987.

MUNNS, R. Comparative physiology of salt and water stress. **Plant and Cell Enviroment,** v.25, p.239-250, 2002.

NAKAMURA, Y.; TANAKA, K.; OHTA, E.; SAKATA, M. Protective effects of external Ca^{2+} on elongation and the intracellular concentration of K^+ in intact mung bean roots under high NaCl stress. **Plant Cell Physiol.,** v.31, p.815- 821, 1990.

NAVARRO, J.M., BOTELLA, M.A., CERDÁ, A., MARTINEZ, V. Phosphorus uptake and translocation in salt-stressed melon plants. **Journal Plant Physiology,** v.158, p.375-381, 2001.

NIEMAN, R.H.; CLARK, R.A. Interactive effects of salinity and phosphorus nutrition on the mature photosynthesising corn leaves. **Plant Physiol.,** v.57, p. 157-161, 1976.

NIEMAN, R.H.; SHANNON, M.C. Screening for salt tolerance. **A.I.D. Joint**

Invitational Workshop on Adaptation of Plant to Mineral Stress in Problem Soils, U.S.D.A., A.R.S., Cornell University, p.2-27. 1977.

NETONDO, G.W.; ONYANGO, J.C.; BECK, E. Sorghum and salinity: II. Gas exchange and chlorophyll fluorescence of sorghum under salt stress. **Crop Science,** v.44, p.806-811,2004.

NOBLE, C.L.; ROGERS, M.J.E. Response of temperate forage legumes to waterlogging and salinity. In: **Handbook of Plant and Crop Stress.** M. Pessarakli ed. Mareei Dekker, Inc. New York. p. 473-496, 1994.

O'LEARY, J.W. Physiological basis for plant growth inhibition due to salinity. In: **Food, fibre, and the arid lands.** McGINNIES, W.G.; GOLDMAN, B.J.; PAYCORE, P. (eds.), Arizona, AASS & Univ. of Arizona Press, p.331-336. 1971.

O'LEARY, J.W. High humidity overcomes lethal levels of salinity in hydroponically grown salt-sensitive plants. **Plant and Soil,** v.42, p.717-721, 1975.

OLIVEIRA, M. Genesis, classification and extension of soils affected by salts. In: GHEYI, H.R.; QUEIROZ, J.E.; MEDEIROS, J.M. (eds) **Manejo e Controle da salinidade na agricultura irrigada.** Campina Grande: UFPB- SBEA, 1997. Chap. 1, p.1-36.

OLIVEIRA, I. P.; CARVALHO, A. M. The caupi crop in the climate and soil conditions of the humid and semi-arid tropics of Brazil. In: ARAÚJO, J. P. P.; WATT, E. E. (Org.) **O caupi no Brasil.** Brasília: IITA; EMBRAPA,. 1988. Chap. 3, p.63-96.

OLIVEIRA, M.; MAIA, C.E. **Physical-chemical quality of water for irrigation in different aquifers in the sedimentary area of the state of Rio Grande do Norte.** Campina Grande, DEAg/UFPB, v.2, n.1, p. 17-21, 1998.

OLIVEIRA, H.P.M. **Effects of iso-osmotic salt concentrations (NaCl + KCl) on growth and ion accumulation in forage sorghum plants,** Fortaleza, CE/UFC, 2002, 32p. Monograph (Agronomy Course, Federal University of Ceará)

ORTIZ, A.; MARTINEZ, V.; CERDÁ, A. Effects of osmotic shock and calcium on growth and solute composition of *Phaseolus vulgaris* plants. Physiol. Plant, v.91, p.468-176, 1994.

PEREIRA, J.R.; VALDIVIESO, C.R.; CORDEIRO, G.G. Recovery of soils affected by sodium through the use of gypsum. In: **SEMINAR ON THE USE OF PHOSPHOGESUM IN AGRICULTURE**, Brasilia, 1985. Brasília, p. 85-105.

PEYRANO, G.; TALEISNIK, E. QUIROGA, M.; FORCHETTI, S.M.; TIGIER, H. Salinity effects on hydraulic conductance, lignin content and peroxidase activity in tomato roots. **Plant Physiology and Biochemistry,** v.35, p.387- 393, 1997.

PONNAMPERUMA, F.N. Role of cultivar tolerance in increasing rice production on saline lands. In: **"Salinity tolerance in plants. Strategies for crop improvement".** STAPLES, R.C.; TOENNIESSEN, G.H. (eds.), John Wiley & Sons, p.257-271. 1984.
QUEIROZ, J.E.; GONÇALVEZ, A.C.; SOUTO, J.S.; FOLEGATTI, M.V. Evaluation and monitoring of soil salinity. In: GHEYI, H.R.; QUEIROZ, J.E.; MEDEIROS, J.F. (Eds) **Management and control of salinity in irrigated agriculture.** Campina Grande: UFPB-SBEA, 1997. Chap. 3. p. 69- 111.

REID, R.J.; SMITH, A. The limits of sodium/calcium interactions in plant growth. **Australian Journal of Plant Physiology,** v.27, p.709-715, 2000.

RHOADES, J.D.; LOVEDAY, J. Salinity in irrigated agriculture. In: STEWART, B.A.; NIELSEN, D.R. (co-eds.) **Irrigation of agricultural crops.** Madison: Amer. Soc. Agron., 1990. chap. 9, p.31-67. (Agronomy, 30)

RHOADES, J.P.; KANDIAH, A.; MASHALI, A.M. **The use saline waters for crop production.** Trad.: GHEYI, H.R.; SOUSA, J.R.; QUEIROZ, J.E. Campina Grande: UFPB, 2000. 117p.

RICHARDS, L. A. **Diagnosis and improvement of saline and alkali soils.** Washington D.C., U.S. Salinity Laboratory, 1954. 160p. (USDA Agriculture Handbook, 60)

ROBERTS, J. K. M.; LINKER, C. S.; BENOIT, A. G.; JARDETZKY, O.; NIEMAN, R H. Salt stimulation of phosphate uptake in maize root tips studies by ^{31}P nuclear magnetic resonance. **Plant Physiology,** v.75, p.947-950, 1984.

ROYO, A.; ARAGÚES, R. Efectos de la salinidad sobre las plantas: tolerância, manejo agronomico, genética y mejora. **Diputación General de** Aragon. 59p. 1989.

RUIZ, D.; MARTINEZ, V.; CERDÁ, A. Citrus response to salinity: growth and nutrient uptake. **Tree Physiology,** v.17, p. 141-150, 1997.

SALAMA, S.; TRIVEDI, S.; BUSHEVA, M.; ARAFA, A. A.; GARAB, G.; ERDEI, L. Effects of NaCl salinity on growth, cation accumulation, chloroplast structure and function in wheat cultivars differing in sail tolerance. J. **Plant Physiol.,** v.144, p.241-247, 1994.

SANTOS, J.G.R. **Salinity in irrigated agriculture: Theory and Practice.** Campina Grande: [s. n.], 2000. 171 p.

SANTOS, R.V.; HERNANDEZ, F.F.F. Recovery of soils affected by salts. In: GHEYI, H.R.; QUEIROZ, J.E.; MEDEIROS, J.M. (eds.). **Management and Control of Salinity in Irrigated Agriculture.** Campina Grande: UFPB- SBEA, 1997. Chap. 10, p.319-361.

SATTI, S. M. E.; IBRAHIM, A. A.; ALKINDI, S. M. Enhancement of salinity tolerance in tomato - implicitons of potassium and calcium in flowering and yield. **Communications in Soil Science and Plant Analysis,** v.25, p.2825- 2840, 1994.

SCHACHTMAN, D.; LIU, W. Molecular pieces to the puzzle of the interaction between potassium and sodium uptake in plants. **Trends Plant Science,** v.4, p.281-287, 1999.

SENA, A.E.S.; BISERRA, J.V. **Efficiencies and economics of sprinkler irrigation in the cultivation of string beans Vigna unguiculata (L.) Walp.** Scientific Production of the PDCT/NE in Ceará, v.2, p.588-605, 1991.

SERRANO, R.; GAXIOLA, R. Microbial models and salt stress tolerance in plants. **Criticai Rev. Plant Sei.,** v.13, p.121-138, 1994.

SERRANO, R. Salt tolerance in plants and microorganisms: toxicity targets and defence responses. **Int. Rev. Cytol.,** v.165, p. 1-52, 1996.

SHANNON, M.C. The effects of salinity on cellular and biochemical processes associated with salt tolerance in tropical plants. In: **Proceedings in plant stress in the tropical environment** (T.L. Davenport & H.M. Harrington, eds.). University of Florida, Kailu-Kona, 1992. p.56-63.

SILVA, J.V. **Effects of CaCh on growth and accumulation of osmoregulators in string bean plants cv. Pitiúba stressed with NaCl.** Fortaleza, UFC, 103p. 1998. (Master's dissertation in Phytotechnics)

SILVA, J.V.; LACERDA, C.F. COSTA, P.H.A.; ENÉAS FILHO, J.; GOMES FILHO, E.; PRISCO, J.T. Physiological responses of NaCl stressed cowpea plants grown in nutrient solution supplemented with CaCh. **Braz. Journal Plant Physiology,** v.15, n.2, p.99-105, 2003.

SILVA JÚNIOR, L.G.A.; GHEYI, H.R.; MEDEIROS, J.F. Composição química de águas do cristalino do Nordeste Brasileiro. **Revista Brasileira de Engenharia Agrícola e Ambiental**, Campina Grande, PB, v.3, n.1, p.11-17, 1999.

SLATYER, R.O. Effects of several osmotic substrates on water relations of tomato. **Aust. J. Biol. Sei.**, v.14, p.519-540, 1961.

SLATYER, R.O. **Plant-water relationships.** Academic Press, Inc. London. 366p. 1967.

SMEDEMA, L.K.; RYCROFT, D.W. **Land drainage: Planning and design of agricultural drainage systems.** London: Bastfort, 1988. 376p.

SOMERO, G.N. Adapting to water stress: convergence on common Solutions. In: **Water and Life.** SOMERO, G.N., OSMOND, C.B., BOLIS, C.L. (eds). Springer-Verlag, Berlin-Heidelberg. Germany, 1992. p. 3-18.

SONG, J.Q.; FUJIYAMA, H. Ameliorative effect of potassium on rice and tomato subjected to sodium salinisation. **Soil Sei. Plant Nutr.**, v.42, n.3, p.493-501, 1996a.

SONG, J.Q.; FUJIYAMA, H. Difference in response of rice and tomato subjected to sodium salinisation to the addition of calcium. **Soil Sei. Plant. Nutr.**, v.42,n.3, p.503-510, 1996b.

SOUSA, M.R. **Behaviour of bean *(Phaseolus vulgaris* L. cv. Eriparza) subjected to different levels of irrigation water salinity.** Lavras :UFLA, 1995. 94p. Master's dissertation in Agricultural Engineering

STROGONOV, B.P. **Physiological basis of salt tolerance of plants - as affected by various types of salinity.** Israel Programme Scientific Translations Ltd. Jerusalem. 279p. 1964.

TAIZ, L.; ZEIGER, E. **Plant Physiology.** 3rd ed. Massachusetts: Sinauer Associates. 690p. 2002.

TERRY, N.; WALDRON, L.J. Salinity, photosynthesis, and leaf growth. **California Agriculture,** v.36, n.10, p.38-39, 1984.

ULZURRUN, M.D.D. Soil salinity and alkalinity in relation to drainage. In: **V MÁSTER INTERNACIONAL DE RIEGO Y DRENAJE.** CENTER-MAPYA, Madrid. 161 p. 2000. Handout

VASCONCELOS, I. MENDES FILHO, P.F.; ALMEIDA, R.T. Nodulation and development of cowpea, *Vigna unguiculata* (L.) Walp., in soils with different levels of salinity. **Revista de Ciência Agronómica**, Fortaleza, v.18, n.1, p. 125-130, 1987.

YANG, Y.W.; NEWTON, R.J.; MILLER, F.R. Salinity tolerance in Sorghum. I. Whole plant response to sodium chloride in S. bicolor and S. halepense. **Crop Science,** v.30, p.775-781, 1990.

YEMM, E. W.; COCKING, E. C. The determination of amino-acids with ninhydrin. **Analyst,** v. 80, p. 209-213, 1955.

YEO, A.R. Salt tolerance in the halophyte *Suaeda marítima* L. Dum: intracellular compartimentation of ions. **Journal of Experimental Botany,** v.32, p.487-497, 1981.

YEO, A. Predicting the interaction between the effects of salinity and climate change on crop plants. **Scientia Horticulturae,** v.78, p. 159-174, 1999.

ZEKRI, M. Effects of NaCl on growth and physiology of sour orange and Cleopatra mandarin seedlings. **Scientia Horticulturae,** v. 7, p.305-315, 1991.

ZEKRI, M.; PARSONS, L.R. Calcium influences growth and leaf mineral concentration of citrus under saline conditions. **Hort. Sei.,** v.25, n.7, p.784- 786, 1990

ZHONG, H. L; LÃUCHLI, A. Spatial distribution of solutes, K, Na, Ca and their deposition rates in the growth zone of primary cotton roots: Effects of NaCl and $CaCl_2$. Planta, v.194, p.34-41, 1994.

ZHU, J.K., HASEGAWA, B.; BRESSAN, R.A. Molecular aspects of osmotic stress in plants. **Crit. Rev. Plant. Sei.,** v.16, p. 253-277, 1997.

MIX
Papier aus verantwortungsvollen Quellen
Paper from responsible sources
FSC® C105338

Printed by Books on Demand GmbH, Norderstedt / Germany